MOLECULAR AND CELL BIOLOGY OF THE PLANT CELL CYCLE

Molecular and Cell Biology of the Plant Cell Cycle

*Proceedings of a meeting held at Lancaster University,
9-10th April, 1992*

Edited by

J.C. ORMROD
I.C.I. Agrochemicals, Jealott's Hill Research Station, Bracknell, U.K.

and

D. FRANCIS
University of Wales, College of Cardiff, U.K.

KLUWER ACADEMIC PUBLISHERS
DORDRECHT / BOSTON / LONDON

Library of Congress Cataloging-in-Publication Data

```
Molecular and cell biology of the plant cell cycle : proceedings of a
  meeting held at Lancaster University, 9-10 April 1992 / edited by
  John Ormrod and Dennis Francis.
      p.   cm.
  Includes index.
  ISBN 0-7923-1767-X (alk. paper)
    1. Plant cell cycle--Congresses.   2. Plant molecular biology-
-Congresses.   3. Plant cellular control mechanisms--Congresses.
  I. Ormrod, John, 1960-   .  II. Francis, D. (Dennis)
  QK725.M74  1993
  581.87'623--dc20                                   92-36385
```

ISBN 0-7923-1767-X

Published by Kluwer Academic Publishers,
P.O. Box 17, 3300 AA Dordrecht, The Netherlands.

Kluwer Academic Publishers incorporates
the publishing programmes of Martinus Nijhoff,
Dr W. Junk, D. Reidel, and MTP Press.

Sold and distributed in the U.S.A. and Canada
by Kluwer Academic Publishers,
101 Philip Drive, Norwell, MA 02061, U.S.A.

In all other countries, sold and distributed
by Kluwer Academic Publishers Group,
P.O. Box 322, 3000 AH Dordrecht, The Netherlands.

printed on acid-free paper

Printed in the Netherlands

Table of contents

Preface

This volume is the published proceedings of a symposium on the regulation of the cell cycle in plant development which formed part of the Spring Meeting of the Society for Experimental Biology (SEB) held in April 1992 at the University of Lancaster, UK. The symposium was sponsored by the Cell Cycle Group of the Society for Experimental Biology and by The British Society for Plant Growth Regulation.

Eight years have elapsed since the previous SEB sponsored publication on cell division in higher plants [1]. Since then cell cycle research has been revolutionised by molecular techniques used to study cell division cycle mutants, notably in the yeasts. More recently, papers have appeared analysing the molecular biology of the cell cycle in plants. In addition, much recent literature on the plant cell cycle has benefited from the high resolution of fluorescence microscopy allied to the detection of antibodies by *in situ* techniques. Therefore, we considered it timely to assemble many of the leading international practitioners of cell and molecular research of the plant cell cycle. We thank each of the contributors and the audience for contributing so enthusiastically to the meeting. Our aim in this volume was to capture the enthusiasm of that meeting and in-so-doing present the current state of the art for studying cell division in plants. We asked our authors to submit either research papers or short reviews which reflected their verbal presentations; we thank them for responding so well. The reader may find, therefore, some overlap if working through the volume from cover to cover.

The volume begins with an overview of check-points of the cell cycle followed by two papers on the genes which regulate the onset of mitosis in plants. There then follows five papers on various aspects of DNA replication followed by two papers which consider the effects of signalling molecules on the transition from G1- to S-phase and studies on mRNAs which are expressed during this transition. Completing this section is an overview of the effect of polyamines on the cell cycle. Structural elements of cell division are considered through a review of meiotic mutants and finally, three overview papers consider the cell cycle in relation to plant development both *in vitro* and *in vivo*.

We hope that many of the ideas and data presented in this volume will act as a spring-board for the characterisation of many more plant homologues to defined cell division cycle genes and that we move nearer understanding the role of cell division cycle genes in meristems and how such activity affects plant development.

JOHN C. ORMROD and DENNIS FRANCIS

References

1. Bryant J.A. and Francis D (1985) The cell division cycle in plants. Cambridge: Cambridge University Press.

1. The controls acting at mitosis in
Schizosaccharomyces pombe

JACQUELINE HAYLES and PAUL NURSE

Abstract

The control over entry into mitosis in many different organisms involves the activity of a key enzyme, p34^{cdc2} which is a serine/threonine protein kinase. In fission yeast p34^{cdc2} activity is regulated by phosphorylation on tyrosine 15 (Y15). Dephosphorylation of Y15 activates the p34^{cdc2} kinase and this initiates mitosis. Regulation of p34^{cdc2} activity requires the function of *cdc*13, *cdc*25, *wee*1 and *mik*1 gene products. The *cdc*13 gene encodes a B type cyclin which forms a complex with p34^{cdc2} and is required for activation of the kinase. The *cdc*25, *wee*1 and *mik*1 genes function on two separate pathways that either activate or inhibit p34^{cdc2} activity. Inactivation of p34^{cdc2} and exit from mitosis involves the destruction of cyclin B and probably also requires the activity of several other gene products including p13^{suc1} and phosphatases. The checkpoint control that ensures that cells do not enter mitosis until DNA replication is completed also acts through p34^{cdc2}.

1. Introduction

During the last few years there has been a considerable increase in our understanding of the eukaryotic cell cycle and in particular the mechanism regulating entry into mitosis [36]. The cell cycle is a highly controlled process that ensures the duplication and accurate segregation of the genetic material to the daughter cells. Several control steps have been identified. There are two major cell cycle controls. The first, called START acts in late G1 and is the point where cells become committed to the mitotic cell cycle as opposed to alternative developmental fates. The second major control step, which we will discuss in more detail later, occurs in G2 and regulates the timing of entry into mitosis. A further control point may act over the exit from mitosis, although this is only poorly understood. There are also dependencies or checkpoints ensuring that cells do not enter mitosis unless DNA replication is completed or do not initiate DNA replication before nuclear division.

   The *cdc*2 gene appears to be a key element in these control mechanisms. In yeast the *cdc*2 gene function has been shown to be required in G1 at START, in G2 at the mitotic control [37] and for the dependencies between M and S [8, 12]. Its function at mitosis is well established but its role in the other control

J.C. Ormrod and D. Francis (eds.), Molecular and Cell Biology of the Plant Cell Cycle, 1–7.
© *1993 Kluwer Academic Publishers. Printed in the Netherlands.*

mechanisms is less well understood. It has been conserved throughout eukaryotic evolution and *cdc2* functional homologues have been found in such divergent organisms as humans, plants and yeast (see PCL John *et al.*, FZ Watts *et al.* and D Dudits *et al.* this volume). The *cdc2* gene encodes a 34kD phosphoprotein, p34[cdc2], which has serine/threonine kinase activity against a range of *in vitro* substrates [45].

2. Regulation of p34[cdc2] kinase activity by phosphorylation

In differentiated cells in both plants and animals the level of p34[cdc2] is low and may be under transcriptional regulation [1, 24, 29]. However in actively dividing cells the level of the *cdc2* transcript and of p34[cdc2] changes very little during the cell cycle [4, 45]. This is in contrast to p34[cdc2] H1 kinase activity which is periodic, peaking at mitosis [31]. In *S. pombe* using histone H1 as a substrate, there is no evidence for p34[cdc2] kinase activity at G1/S. However, in other systems a *cdc2*-like kinase activity has been reported to be correlated with S phase and a partially purified activity, RF-S, which promotes DNA replication in G1 cell extracts contains p34[cdc2] [9].

The activity of p34[cdc2] at mitosis is regulated by phosphorylation. The level of phosphorylation decreases as cells enter mitosis but it does not disappear completely [17]. The amino acid residues that are phosphorylated have been mapped in several different organisms. In *S. pombe* p34[cdc2] is phosphorylated on tyrosine 15 (Y15) and threonine 167 (T167) [21]. Y15 lies within the consensus sequence for nucleotide binding GXXGXG. Phosphorylation on this residue is inhibitory for entry into mitosis and for kinase activation and may act by blocking the binding of ATP. In the fission yeast, p34[cdc2] is activated by dephosphorylation of Y15 [21].

In higher eukaryotes the situation may be more complex as phosphorylation also occurs on threonine 14 (T14) in the ATP binding site [25, 34]. Dephosphorylation of both T14 and Y15 must occur before p34[cdc2] kinase is activated [34]. In *Saccharomyces cerevisiae* p34[cdc28] (the p34[cdc2] homologue) is phosphorylated on the equivalent residue to Y15 (Y19) [3, 46]. The phosphorylation state of Y19 does not seem to regulate entry into mitosis or be involved in the dependency of M on the completion of DNA replication, as we shall discuss later [3, 46]. *S. cerevisiae* has an unusual cell cycle in that mitosis is initiated soon after completion of start, around the time of DNA replication. The cells have a short spindle present for most of the cell cycle before elongation of the spindle and the completion of mitosis is initiated [35]. Therefore it might be expected that a control acting over the initiation of mitotic continuation may be a different mechanism to that over the entry into mitosis.

Threonine phosphorylation of p34[cdc2] also occurs on residue T167 (in *S. pombe*) [20] and T161 (in mammals) [25, 34]. This site of phosphorylation is essential for kinase activity. It has also been reported that chick p34[cdc2] is phosphorylated on serine 283 during G1 [25]. This does not appear to be the case for fission yeast p34[cdc2] [16].

3. Gene functions regulating tyrosine phosphorylation of p34[cdc2]

In fission yeast a phosphorylation network has been demonstrated to act over the regulation of the timing of mitosis. The *wee1, nim*1 and *mik*1 genes all encode protein kinases and act through p34[cdc2] to inhibit entry into mitosis [30, 43, 44]. The *cdc*25 acts on a separate pathway to activate p34[cdc2] [42] and has been shown to directly dephosphorylate p34[cdc2] *in vitro* [26, 47]. These inhibitory and activatory pathways may act to regulate the level of tyrosine phosphorylation of p34[cdc2] and hence the timing of entry into mitosis.

A temperature sensitive (ts) *cdc*25-22 mutant at the restrictive temperature will block in late G2 just before mitosis. These cells accumulate tyrosine phosphorylated p34[cdc2] [21]. Lack of *wee*1 function is not lethal but causes cells to be advanced into mitosis at a small cell size [38]. In this situation p34[cdc2] still has detectable Y15 phosphorylation. A deletion of *mik*1 does not advance cells into mitosis, but absence of both the *wee*1 and *mik*1 gene functions advances cells into mitosis prematurely causing a mitotic catastrophe and cell death [30]. In this situation, p34[cdc2] is completely dephosphorylated on Y15. These data show that although *wee*1 and *mik*1 probably have overlapping functions and both affect the level of Y15 phosphorylation of p34[cdc2], the *wee*1 function has the greater effect on the timing of entry into mitosis. The level of *wee*1 protein, p107[wee1], does not appear to vary during the cell cycle whereas the level of p80[cdc25] shows a gradual increase and decrease, being maximal at mitosis [15, 33].

4. p34[cdc2] interaction with cyclin B is required for kinase activity

Cyclins A and B were originally identified as proteins synthesised during the cell cycle and destroyed at each mitosis [14]. Since that time several different classes of cyclins have been identified and the definition of a cyclin and the cyclin consensus sequence have now become rather hazy. However, the role of cyclin B at mitosis is quite well established. Maturating Promoting Factor (MPF) has now been purified from several different organisms and consists of a complex between p34[cdc2] and cyclin B [11, 18, 19, 27, 28]. In fission yeast, cyclin B is encoded by the *cdc*13 gene and is essential for entry into mitosis [6, 22]. Like other cyclins it is degraded during mitosis, at the metaphase/anaphase transition [6, 31]. Although p56[cdc13] is required for activation of the p34[cdc2] kinase it has not yet been established whether the complex is needed after activation of the kinase. In *S. pombe* the H1 kinase appears still to be active after degradation of cyclin B and it is possible that other elements are also required for inactivation of the kinase at the end of mitosis [31]. The p34[cdc2] / cyclin B complex is probably present for most of S-phase and G2, in the inactive form [10]. During the vegetative cell cycle it is the level of Y15 phosphorylation on p34^{cdc2} and not the accumulation of cyclin that regulates the timing of entry into mitosis under normal growth conditions [21].

5. Regulation of the exit from mitosis

As described in the previous section p56^{cdc13} degradation at the metaphase/anaphase transition precedes the loss of p34^{cdc2} activity. This suggests that other gene functions may be required for inactivation of p34^{cdc2} at the end of metaphase. One such gene product is p13^{suc1} [7]. This gene was first identified as an allele-specifie supressor of cdc2^{ts} mutants [23] and MPF has been purified by the ability of p13^{suc1} to bind strongly to p34^{cdc2} *in vitro* [27, 28].

Cells carrying a deletion of *suc*1 become blocked in late mitosis with activated p34^{cdc2} [31]. This results suggests that p13^{suc1} may be involved in the process leading to inactivation of p34^{cdc2} at the end of mitosis. As well as inactivation of p34^{cdc2}, phosphatases may be required for progress through and exit from mitosis. In *S. pombe* the activity of type 1 phosphatases encoded by *dis*2/*bws*1 gene and *sds*21 are required at the later stages of mitosis [39]. A further gene *sds*22 which may modulate type 1 phosphatases is also required for the metaphase/anaphase transition [40]. These gene functions may be necessary to dephosphorylate key substrates that have been phosphorylated during mitosis. This may be necessary before cells can exit from mitosis; both *suc*1 and *dis*2/*bws*1 may also be required for entry into mitosis [5]. Overexpression of *dis*2/*bws*1 suppresses the ability of a *cdc*25-22 *wee*1-50 double mutant to enter mitosis. Overproduction of *suc*1 in wild type cells also delays entry into mitosis [23]. In both cases overproduction of a gene whose function is required for exit from mitosis may also prevent entry into mitosis.

6. Signals affecting the timing of entry into mitosis

There is a dependency or checkpoint that ensures the completion of DNA replication before cells enter mitosis. In wild type cells if DNA replication is incomplete the cells will not enter mitosis. Wee mutants of *S. pombe* are advanced into mitosis at a small cell size [13]. Of these mutants some were able to enter mitosis even when DNA replication was blocked in the presence of hydroxyurea (HU). One of these mutants was *cdc*2-3*w* which suggests that the signal for incomplete replication of DNA is coupled to the mitotic control through *cdc*2. The other mutants that could bypass this check point were *cdc*-F15, cells overexpressing *cdc*25 [13] or cells lacking both *wee*1 and *mik*1 [30]. In each of these situations there is a loss of Y15 phosphorylation. This loss of Y15 phosphorylation is correlated with the loss of the checkpoint control, suggesting that it may mediate the checkpoint signal. The *cdc*25 and *wee*1 gene functions regulate the level of *cdc*2 Y15 phosphorylation and can influence the checkpoint control. These two genes probably do not directly act on the checkpoint pathway as loss of both activities does not abolish the checkpoint [12].

To investigate how the unreplicated DNA generates a signal and how this is coupled to the mitotic control, mutants defective in this checkpoint have been isolated. These fall into eight linkage groups. Three of the groups are allelic to previously identified radiation sensitive (*rad*) genes, *rad*1, 3 and 17. The other groups represented novel HU sensitive (*hus*) genes, *hus*1-5. All of these hus mutants are sensitive to radiation damage (T Enoch, A Carr, P Nurse, in press). Analysis of the *rad* mutants has shown that as well as *rad* 1, 3 and 17, a further gene, *rad*9, is also defective in this checkpoint control [2,41]. It seems likely that the signal from unreplicated DNA and damaged DNA feed into the same pathway which requires activities of the *rad* and *hus* genes. This pathway probably acts through p34^{cdc2} to delay mitotic onset. These genes are at present being cloned and further analysis should be useful for understanding how this checkpoint functions.

7. Conclusion

Many of the elements required during mitosis both for regulation of entry into and exit from mitosis have been conserved throughout evolution. It is therefore highly likely that the genetic analysis in fission yeast will provide a useful basis for understanding the cell cycle in such divergent organisms as plants and humans.

Acknowledgements

We thank Chris Norbury for reading the manuscript and the Imperial Cancer Research Fund for financial support.

References

1. Akhurst RJ, Flavin NB, Worden J and Lee MG (1989) Intracellular localisation and expression of mammalian CDC2 protein during myogenic differentiation. Differentiation 40: 36-41.
2. Al-Khodairy F and Carr AM (1992) Mutants defining the G2 checkpoint pathway in S. *pombe*. EMBO J 11: 1343-1350.
3. Amon A, Surana U, Muroff I and Nasmyth K (1992) Regulation of p34^{cdc28} tyrosine phosphorylation is not required for entry into mitosis in S. *cerevisiae*. Nature 355: 368-371.
4. Aves SJ, Durkacz BW, Carr A and Nurse P (1985) Cloning, sequencing and transcriptional control of the *Schizosaccharomyces pombe cdc*10 'start' gene. EMBO J 4: 457-463.
5. Booher R and Beach D (1989) Involvement of a type 1 protein phosphatase encoded by *bws*1 + in fission yeast mitotic control. Cell 57: 1009-1016.
6. Booher RN, Alfa CE, Hyams JS and Beach DH (1989) The fission yeast cdc2/cdc13/suc1

6

protein kinase: regulation of catalytic activity and nuclear localization. Cell 58: 485-497.

7. Brizuela L, Draetta G and Beach D (1987) p13[suc1] acts in fission yeast cell cycle as a component of the p34[cdc2] protein kinase. EMBO J 6: 3507-3514.

8. Broek D, Bartlett R, Crawford K and Nurse P (1991) Involvement of p34[cdc2] in establishing the dependency of S phase on mitosis. Nature 349: 388-393.

9. D'Urso G, Marraccino RL, Marshak DR and Roberts JM (1990) Cell cycle control of DNA replication by a homologue from human cells of the p34[cdc2] protein kinase. Science 250: 786-791.

10. Draetta G and Beach D (1988) Activation of cdc2 protein kinase during mitosis in human cells: cell cycle-dependent phosphorylation and subunit rearrangement. Cell 54: 17-26.

11. Dunphy WG, Brizuela L, Beach D and Newport J (1988) The Xenopus homolog of cdc2 is a component of MPF, a cytoplasmic regulator of mitosis. Cell 54: 423-431.

12. Enoch T, Gould K and Nurse P (1992) Mitotic checkpoint control in fission yeast. In: Cold Spring Harbor Symp Quant Biol, 56: 409-416.

13. Enoch T and Nurse P (1990) Mutation of fission yeast cell cycle control genes abolishes dependence of mitosis on DNA replication. Cell 60: 665-673.

14. Evans T, Rosenthal E, Youngbloom J, Distel D and Hunt T (1983) Cyclin: a protein specified by maternal mRNA in sea urchin eggs that is destroyed at each cleavage division. Cell 33: 389-396.

15. Featherstone C and Russell P (1991) Fission yeast p107[wee1] mitotic inhibitor is a tyrosine/serine kinase. Nature 349: 808-811.

16. Fleig UN and Nurse P (1991) Expression of a dominant negative allele of cdc2 prevents activation of the endogenous p34[cdc2] kinase. Mol Gen Genet 226: 432-440.

17. Gautier J, Matsukawa T, Nurse P and Maller J (1989) Dephosphorylation and activation of Xenopus p34[cdc2] protein kinase during the cell cycle. Nature 339: 626-629.

18. Gautier J, Minshull J, Lohka M, Glotzer M, Hunt T and Maller JL (1990) Cyclin is a component of MPF from Xenopus. Cell 60: 487-494.

19. Gautier J, Norbury C, Lohka M, Nurse P and Maller J (1988) Purified maturation-promoting factor contains the product of a Xenopus homolog of the fission yeast cell cycle control gene cdc2+. Cell 54: 433-439.

20. Gould K, Moreno S, Owen D, Sazer S and Nurse P (1991) Phosphorylation at Thr 167 is required for fission yeast p34[cdc2] function. EMBO J 10: 3297-3309.

21. Gould KL and Nurse P (1989) Tyrosine phosphorylation of the fission yeast cdc2+ protein kinase regulates entry into mitosis. Nature 342: 39-45.

22. Hagan I, Hayles J and Nurse P (1988) Cloning and sequencing of the cyclin-related cdc13+ gene and a cytological study of its role in fission yeast mitosis. J. Cell Sci 91: 587-595.

23. Hayles J, Aves S. and Nurse P (1986) suc1 is an essential gene involved in both the cell cycle and growth in fission yeast. EMBO J 5: 3373-3379.

24. John PCL, Sek FJ, Carmichael JP and McCurdy DW (1990) p34[cdc2] homologue level, cell division, phytohormone responsiveness and cell differentiation in wheat leaves. J Cell Sci 97: 627-630.

25. Krek W and Nigg EA (1991) Differential phosphorylation of vertebrate p34[cdc2] kinase at the G1/S and G2/M transitions of the cell cycle: identification of major phosphorylation sites. EMBO J 10: 305-316.

26. Kumagai A and Dunphy WG (1991) The cdc25 protein controls tyrosine dephosphorylation of the cdc2 protein in a cell-free system. Cell 64: 903-914.

27. Labbé JC, Capony JP, Caput D, Cavadore JC, Derancourt MK, Lelias JM, Picard A and Dorée M (1989) MPF from starfish oocytes at first meiotic metaphase is a heterodimer containing one molecule of cdc2 and one molecule of cyclin B. EMBO J 8: 3053-3058.

28. Labbé JC, Picard A, Peaucellier G, Cavadore JC, Nurse P and Dorée M (1989) Purification of MPF from starfish: identification as the H1 histone kinase p34[cdc2] and a possible mechanism for its periodic activation. Cell 57: 253-263.

29. Lee MG, Norbury CJ, Spurr NK and Nurse P (1988) Regulated expression and phosphorylation

of a possible mammalian cell-cycle control protein. Nature 333: 676-679.

30. Lundgren K, Walworth N, Booher R, Dembski M, Kirschner M and Beach D (1991) *mik*1 and *wee*1 cooperate in the inhibitory tyrosine phosphorylation of *cdc*2. Cell 64: 1111-1122.

31. Moreno S, Hayles J and Nurse P (1989) Regulation of p34[cdc2] protein kinase during mitosis. Cell 58: 361-372.

32. Moreno S and Nurse P (1991) Clues to action of cdc25 protein. Nature 351: 194.

33. Moreno S, Nurse P and Russell P (1990) Regulation of mitosis by cyclic accumulation of p80[cdc25] mitotic inducer in fission yeast. Nature 344: 549-552.

34. Norbury C, Blow J and Nurse P (1991) Regulatory phosphorylation of the p34[cdc2] protein kinase in vertebrates. EMBO J 10: 3321-3329.

35. Nurse P (1985) Cell cycle control genes in yeast. Trends in Genetics 1: 51-55.

36. Nurse P (1990) Universal control mechanism regulating onset of M-phase. Nature 344: 503-508.

37. Nurse P and Bissett Y (1981) Gene required in G1 for commitment to cell cycle and in G2 for control of mitosis in fission yeast. Nature 292: 558-560.

38. Nurse P and Thuriaux P (1980) Regulatory genes controlling mitosis in the fission yeast Schizosaccharomyces pombe. Genetics 96: 627-637.

39. Ohkura H, Kinoshita N, Miyatani S, Toda T and Yanagida M (1989) The fission yeast *dis*2+ gene required for chromosome disjoining encodes one of two putative type 1 protein phosphatases. Cell 57: 997-1007.

40. Ohkura H and Yanagida M (1991) S. pombe gene *sds*22+ essential for a midmitotic transition encodes a leucine-rich repeat protein that positively modulates protein phosphatase-1. Cell 64: 149-157.

41. Rowley R S S and Young P (1992) Checkpoint controls in Schizosaccharomyces pombe: *rad*1. EMBO J. 11: 1335-1342.

42. Russell P and Nurse P (1986) *cdc*25+ functions as an inducer in the mitotic control of fission yeast. Cell 45: 145-153.

43. Russell P and Nurse P (1987) The mitotic inducer *nim*1+ functions in a regulatory network of protein kinase homologs controlling the initiation of mitosis. Cell 49: 569-576.

44. Russell P and Nurse P (1987) Negative regulation of mitosis by *wee*1+, a gene encoding a protein kinase homolog. Cell 49: 559-567.

45. Simanis V and Nurse P (1986) The cell cycle control gene *cdc*2+ of fission yeast encodes a protein of fission yeast encodes a protein kinase potentially regulated by phosphorylation. Cell 45: 261-268.

46. Sorger PK and Murray AW (1992) S-phase feedback control in budding yeast independent of tyrosine phosphorylation of p34[cdc28]. Nature 355: 365-368.

47. Strausfeld U, Labbé J-C, Fesquet D, Cavadore JC, Picard A, Sadhu K, Russell P and Dorée M (1991) Dephophorylation and activation of a p34[cdc2]/cyclin B complex *in vitro* by human CDC25 protein. Nature 351: 242-245.

2. A p34^{cdc2}-based cell cycle: its significance in monocotyledonous, dicotyledonous and unicellular plants

PETER C.L. JOHN, KERONG ZHANG and CHONGMEI DONG

Abstract

A requirement in proliferating cells for attainment of adequate cell size for progress through the cell cycle provides a means for coordinating rates of division with rates of growth. Control points of the plant cell cycle are in late G1-phase and at the initiation of mitosis. A p34^{cdc2}-based cell cycle in plants has been indicated by changes in the amount, phosphorylation and activity of p34^{cdc2}-like protein in late G1 and at mitosis, and by association of plant p34^{cdc2} protein with two regulatory binding proteins. Plant p34^{cdc2} bound *in vitro* to p13^{suc1} protein, which is necessary for progress through anaphase in fission yeast, and plant p34^{cdc2} co-purified with a cyclin B-like 56 kDa protein that was recognised by antibody against p56^{cdc13}/cyclin B protein of fission yeast. A sharp peak in activity of p34^{cdc2} protein kinase correlated with the transient presence of cyclin B-like protein during normal mitosis. Moreover, in a metaphase-arresting mutant persistence of cyclin B protein correlated with persistence of activated p34^{cdc2} kinase activity and with inability to progress to anaphase under restrictive conditions. Selective inhibition of individual events at mitosis revealed that chromosome condensation, spindle formation, nuclear envelope breakdown and preprophase band disassembly occur independently following an initiating stimulus at prophase that coincides with, and probably derives from, the activation of p34^{cdc2} kinase. In development, the presence of p34^{cdc2}-like protein was restricted to meristem tissues and to stages of organogenesis in which division is appropriate. During cell differentiation in diverse tissues, and probably universally, a more than fifteen fold reduction of p34^{cdc2} established a secure exit from the cell cycle and resumption of division required restoration of p34^{cdc2} levels. Both auxin and cytokinin phytohormones influenced the induced levels of the cell cycle control protein.

1. Introduction

An important attribute of the cell division cycle is that its frequency is coordinated with the rate of growth and therefore cell size is stabilised [16]. Our recent results indicate that plant cell division can be coordinated with growth by a requirement for attainment of a critical minimum cell size before the events termed the 'DNA-division' cycle by Mitchison [49] are initiated. We also

9

J.C. Ormrod and D. Francis (eds.), Molecular and Cell Biology of the Plant Cell Cycle, 9–34.
© 1993 *Kluwer Academic Publishers. Printed in the Netherlands.*

increasingly recognise that key catalysts of this sequence include protein kinase enzymes closely related to the threonine serine protein kinase that is encoded by the cdc2 gene of *Schizosaccharomyces pombe* (see J Hayles and P Nurse, this volume). This enzyme has been unequivocally demonstrated by genetic analysis in the yeasts as the key cell cycle control molecule and as the focus of a network of regulatory proteins (reviewed by P. Nurse [54]). One of these, p13[suc1], which is necessary for completion of mitosis, binds to the plant p34[cdc2]-like enzyme. The plant enzyme also co-purified with a cyclin B-like protein that is recognised by antibody raised against the p56[cdc13]/cyclin B protein of *S. pombe* and the abundance of B cyclin in the plant cell is coupled tightly to the peak of p34[cdc2] enzyme activity.

At the tissue level three aspects of plant cell division control that are essential for the formation of organised tissues are now also better understood. Cell division must be restricted to specialised meristematic regions allowing cells outside to develop structures and functions that often require their enlargement to several multiples of the critical size for initiation of division. We now know that a more than fifteen-fold decline in the level of p34[cdc2]-like proteins in cells leaving the meristem preceeds their subsequent differentiation and, in view of the essential contribution of p34[cdc2] at the start of DNA replication and at mitosis [55, 7], this can be seen to enforce exit from the cell cycle. The second aspect of division in tissues that is now better understood is the mechanism by which the cell cycle can be resumed in the establishment of lateral meristems, secondary thickening, or wound of repair and also during plant regeneration after gene transfer to protoplasts. We now understand that resumption of division by differentiated cells requires a more than fifteen-fold increase in level of p34[cdc2]-like proteins. The third aspect of division control in tissues for which we begin to identify regulatory elements is the control of preprophase band (PPB) disassembly. The PPB microtubules form a ring in the cortical cytoplasm and mark the site at which the new cross wall will fuse with the old. PPBs are cleared from this site when formation of the new wall is imminent at the initiation of metaphase [25]. The PPB is unique to plant cells and its presence indicates the importance of determining the direction in which new cells are formed for the occurrence of organogenesis, which in plants must be carried through without alteration of relative cell positions (see D Francis and RJ Herbert, this volume). Breakdown of PPBs is normally dependent upon the initiation of metaphase and we have found that it correlates with activation of plant p34[cdc2]-like protein kinase activity. We have observed that the mitotic sequence, of chromosome condensation, spindle formation, nuclear envelope and preprophase band (PPB) breakdown, does not derive from each earlier event triggering the next since selective inhibition of spindle formation does not prevent PPB breakdown, rather the events occur independently following a common initiating signal.

2. Control points coordinating division with growth

In meristematic tissue of a growing plant, as in populations of microorganisms, the maintenance of optimium cell size depends upon balancing the rates of cell division and growth. Cells entering division are both older and larger than when they were formed and in principle either their age or their size could trigger division [16]. However, size homeostasis is more rapidly achieved if the start of processes leading to cell division is dependent upon attainment of a critical minimum size. This is difficult to test in higher plant tissues, in which it is difficult to obtain populations of single cells that are genetically normal and are approaching division with sharp synchrony at a range of experimentally determined growth rates. However unicellular green plants can be readily synchronised and we have used such cultures of *Chlamydomonas* growing at different rates, determined by light intensity or limiting CO_2, and found that a critical minimum cell size must be exceeded for G1-phase to be terminated by commitment to carry through the DNA-division processes of S-phase, G2-phase, mitosis and cytokinesis [34]. Growth is necessary for attainment of the necessary cell size and for execution of the commitment process [10] but once division is committed further growth is unnecessary for completion of the DNA-division sequence [9]. The late G1 control point of the unicellular plant has therefore been recognised [33, 34] as equivalent to the START division control of budding yeast [59]. Their functional equivalence includes the following; both initiate a portion of the cell cycle that is of relatively constant duration over a wide range of growth rates and cell cycle durations, indicating that START is the rate limiting control point for division, and in both cases a single doubling of nuclear DNA and cell number is committed and begins within 0.1 cycles. The cell size requirement is indicated by the constancy at different growth rates of cell size at START [45, 34]. The unicellular plant divison cycle allows a clearer test than does the yeast, of whether size is a crucial determinant. This test is possible because *Chlamydomonas* has a requirement for a minimum time in G1-phase, which keeps the growth-dependent G1-phase concurrent with daylight at moderate to fast growth rates. At the higher growth rates the timed extension of G1-phase results in large mother cells but daughter size is stabilised by the proportionate recurrence of additional commitments, each to a DNA and cell doubling, resulting in multiple fission at the end of the cycle. By centrifugally selecting large and small daughter cells and growing them through the cell cycle at the same rate, a higher division number was seen in the larger cells and was unequivocally due to their larger size [10]. Therefore size of a plant cell can be a critical determinant of commitment to division and the equivalence of the late G1 (START) control points, at which cell size exerts an influence in all eukaryotes that have been tested, has been emphasised earlier [33] and reviewed recently [3], is illustrated in Fig. 1.

12

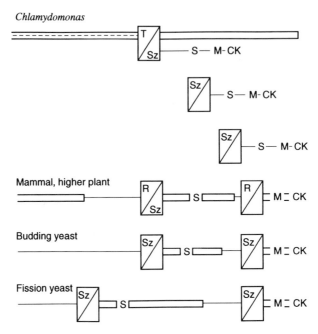

Chlamydomonas

Mammal, higher plant

Budding yeast

Fission yeast

Fig. 1. Universal control points in the cell cycles of plant and other eukaryote cells (indicated by box symbols). The late G1 control point that starts progress to DNA synthesis (S) and is often termed commitment or START (see text) is influenced by cell size in budding yeast [45], fission yeast [56], mammalian cells [2, 45], *Chlamydomonas* [10, 34] and probably higher plant cells. The initiation of mitosis (M) is also influenced by cell size in the yeasts [56, 33]. Key; T, expiry of a critical time; Sz, attainment of a critical size; R, restriction point, potentially influenced by hormone; ===, temperature compensated timer that at higher growth rate extends G1-phase in *Chlamydomonas* and necessitates repetition of the start control at higher growth rate; ——, timed period little influenced by growth rate e.g. S, M and cytokinesis (CK) phases. The similarity of mechanisms at the major control points was first demonstrated by Beach, Durkacz and Nurse [2]. Diagram adapted from John [33].

The multiple fission form of cell cycle can now be understood as simply resulting from repetition of a universal START division motif. In recognition of this it has been suggested that evolution of the multiple fission cycle could have arisen by repetition of the p34^{cdc2} function [33]. Remarkably, fission yeast cells have recently been found to recapitulate this postulated evolutionary step when carrying mutations that cause instability of p34^{cdc2} but not of other proteins. Synthesis of new p34^{cdc2} results in repetition of the START division function and reduplication of DNA within a single cell cycle [7].

Recognition of the importance of cell size and of the START control in plants and evidence of the involvement of p34^{cdc2}-like protein, as will be described below, leads to four propositions. The principal control point hypothesis [70] should be refined [37] to recognise single transitions, at late G1 and at mitotic initiation, as the principal control points. In meristems, the rate of division can be adjusted to accommodate environmentally imposed changes in growth rate by a requirement for a critical minimum size for cell cycle progress [40]. Where differentiation involves cell enlargement a supression of division must occur to allow the alternative developmental programme and this supression of division often involves the establishment of low levels of the key cell cycle catalyst p34^{cdc2}, as will be illustrated below. The principle of selection for maximum economy of mechanism suggests that the sizer mechanism for division control will also set maximum differentiated cell size as a multiple of cell birth size.

3. Presence in plants of p34^{cdc2} and its regulatory binding proteins

The equivalence of the START cell cycle controls (Fig. 1) led us to seek plant molecules that are homologous with those that act at this point in other eukaryotes. The *CDC*28 gene was first identified as essential at START in budding yeast [59] and the unity of eukaryote cell cycles was first demonstrated by substitution of *CDC*28 for *cdc*2 in fission yeast [2] then extended by detection of a human homologue of *cdc*2 [44]. The yeast and human sequences revealed highly conserved sequences for ATP binding and phosphorylation, which indicated protein kinase function, and also regions that were specific for the *cdc*2 cell cycle protein (see J Hayles and P Nurse, this volume). The largest of these was EGVPSTAIREISLLKE (abbreviated as PSTAIR) which has proved to be perfectly conserved in all *cdc*2 homologues. Antibody raised against PSTAIR peptide reacted with 34 kDa protein in unicellular, monocotyledonous and dicotyledonous higher plants [37]. Taxa giving a positive signal included the red algae, which diverged so early in eukaryote evolution that they lack the flagellum and have captured a different photosynthetic prokaryote [64, 32]. This has led to the proposition that a p34^{cdc2}-based mechanism for DNA replication and segregation was an integral part of the evolution of the eukaryote cell and was subsequently inherited in all eukaryotes because it was too essential for subsequent evolutionary change [37].

Biochemical evidence for the presence of p34^{cdc2}-like protein in plants includes: (i) protein kinase activity that co-purifies with PSTAIR-containing proteins and binds to authentic p13^{suc1} ([36] and Figs. 3 and 7), that has activity independent of calcium and cyclic AMP, utilises H1 histone or the peptide based on large T antigen that is phosphorylated by authentic p34^{cdc2} but not six other protein kinases [47, 38]; (ii) presence in the enzyme that is purified from wheat meristem cells, by DEAE cellulose and p13^{suc1} binding and elution, of a cyclin B-like protein of 56 kDa that is recognised on Western blots by antibody against the p56^{cdc13}/cyclin B protein [52] of *S. pombe* (Fig. 2), and (iii) a temporal association of cyclin proteins with mitotic activity in dividing and in genetically blocked cells. This evidence will be presented more fully in relation to the cell cycle involvement of plant p34^{cdc2}. Definitive genetic evidence of homology of some plant p34^{cdc2}-like proteins has come from the ability of part of the pea *cdc*2 gene and entire *cdc*2 genes from *Arabidopsis*,

14

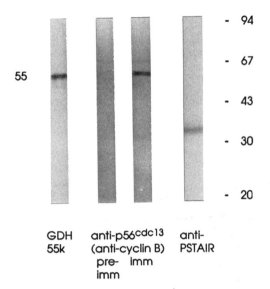

- 94

- 67

55

- 43

- 30

- 20

GDH
55k

anti-p56cdc13
(anti-cyclin B)
pre-   imm
imm

anti-
PSTAIR

Fig. 2. Presence of PSTAIR-containing p34cdc2-like protein and p56cdc13/cyclin B-like protein, in protein kinase enzyme purified from wheat meristem cells by DEAE cellulose ion exchange and p13suc1 affinity column chromatography [36]. The 34 kDa protein was detected by antibody raised against the peptide EGVPSTAIREISLLKE and the 56 kDa protein was detected by immune (imm) but not by pre-immune (pre-imm) antibody raised against the p56cdc13/cyclin B protein purified after overexpression in *E. coli* [52]. The cyclin B protein is larger than the glutamate dehydrogenase (GDH) 55 kDa size marker. The purified fraction that was analysed by Western blotting had calcium and cyclic AMP independent protein kinase activity, assayed as described [36], that phosphorylated H1 histone and a synthetic peptide ADAQHATPPKKKRKVEDPKDF [47] that is preferentially phosphorylated by p34cdc2 but not by six different classes of protein kinase.

maize and alfalfa to complement yeast cells lacking *cdc2* or *CDC28* [8, 17, 30, 31]. It is becoming clear that in higher eukaryotes, which are under less extreme selective pressure to retain the minimum genome size than are the yeasts because speed of nutrient conversion into new cells is a less significant factor in their evolutionary success, there has been a duplication of *cdc2/CDC28*-like genes. Not all of these have retained every function of the single yeast *cdc2* gene, perhaps because some functions are contributed by other members of the family in the same cell. The aquisition of additional functions by these incompletely homologous cyclin-dependent kinases (or *CDK*s) [58], is an area of active current study. All the known variants of p34cdc2 retain the PSTAIR sequence and none have been identified as having other than cell cycle function (reviewed [58]). Antibody against PSTAIR is therefore a good reagent to test for the presence or absence of the p34cdc2 family of protein kinases during developmental changes.

4. Involvement of p34[cdc2] in the plant cell cycle

Several forms of evidence indicate that p34[cdc2] is actively involved in plant cell division (also see D Dudits *et al.*, this volume).

An increase in amount of the p34[cdc2] protein coincides with commitment to division in late G1-phase and the correlation is maintained when extreme nutritional and environmental influence is used to change the timing over a two-fold range [37]. It should be noted that the increase is peculiar to synchronising conditions, in which daughter cells begin G1-phase by emerging from a quiescence imposed by a preceding dark period. Their resumption of the cell cycle is similar to serum stimulation of quiescent mammalian cells, which also show an increase of p34[cdc2] at the late G1 control point [43]. Although this increase indicates the need for p34[cdc2] at this point in the cell cycle the increase is not in itself essential since cells that are not emerging from quiescence have adequate p34[cdc2] and can accumulate it steadily through the cell cycle in balance with other proteins, as we have observed in aphidicolin synchronised cells of *Nicotiana plumbaginifolia* (K Zhang and PCL John, unpublished) and has been observed in malignant human HeLa cells that do not enter quiescence [12].

The phosphorylation pattern of plant p34[cdc2]-like proteins correlates with cell cycle phase [37]. Phosphorylated forms of the enzyme were detected by the reduction in their electrophoretic mobility and by presence of $[^{32}P]$ PO_4 in protein immunoprecipitated by antibody against whole p34[cdc2]. A phosphorylated form of the enzyme first appeared in late G1 at commitment and a hyperphosphorylated form at the beginning of mitosis. A close parallel exists with the behaviour of p34[cdc2] in HeLa cells [12] and the phosphorylated plant and animal proteins had the same mobilities when electrophoresed in parallel. A minimum hypothesis [37] postulated progressive phosphorylation of plant p34[cdc2] beginning in late G1 and becoming more extensive prior to mitosis then declining in late mitosis. This interpretation is consistent with recent detailed evidence from chicken hepatoma and other animal cells [41] that p34[cdc2] is phosphorylated at serine 277 in late G1-phase, at threonine 14 and tyrosine 15 in G2-phase and that the latter two are removed at mitosis concomitant with catalytic activation of p34[cdc2] kinase (see J Hayles and P Nurse, this volume). The limited data from plants must be interpreted with caution because we lack information concerning the sites of phosphorylation in plant p34[cdc2] and cannot eliminate the possibility that variants of the protein are also present in phosphorylated form, although there is no evidence that these are abundant. This is an area for future study in all eukaryotes.

Plant p34[cdc2]-like protein binds to the p13[suc1] protein of fission yeast that is necessary for p34[cdc2] inactivation at anaphase and for completion of mitosis [52]. This binding capacity can be used to affinity purify p34[cdc2], using p13[suc1] in immobilised form and then in free form for eluting [6], and by this means we have purified the p34[cdc2]-like protein from the meristem region of wheat [36]. Affinity for p13[suc1] has also been observed in p34[cdc2]-like proteins from *Chlamydomonas* [38], tobacco and wheat and purification of the enzyme by

16

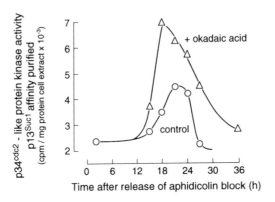

y-axis label: p34^cdc2 - like protein kinase activity
p13^suc1 affinity purified
(cpm / mg protein cell extract × 10⁻³)

x-axis label: Time after release of aphidicolin block (h)

+ okadaic acid

control

Fig. 3. Activity of p34^cdc2-like protein kinase in the cell cycle of *Nicotiana plumbaginifolia* cells in suspension culture were synchronised with aphidicolin and cultured during the sampling period in the presence (△), or absence (○), of 12 μM okadaic acid as described [71]. Before assay p34^cdc2 was purified by p13^suc1 affinity chromatography essentially as described [36] but with additional washes of 1% NP40 buffer before elution with 0.5 mg/ml p13^suc1 and assay as described [36]. The timing of division events in the culture is illustrated in Fig. 4 d, e.

p13^suc1 binding has allowed study of changes in catalytic activity through the cell cycle (Fig. 3) and in development (Fig. 7). The presence of binding capacity in the plant enzyme correlates with the presence of plant homologues of p13^suc1 detected by affinity purified anti-p13^suc1 antibody, in wheat, *Chlamydomonas*, and pea, indicating a probable universal occurrence and leading to the suggestion that the interaction of p34^cdc2 and p13^suc1 is an essential part of mitotic control that has therefore been retained during evolution [36].

Cell cycle related changes in catalytic activity of p34^cdc2 correlate with changes in phosphorylation of p34^cdc2-like proteins, in the unicellular plant. More than ten-fold changes in protein kinase activity, directed to either H1 histone or to a p34^cdc2-preferred peptide [47] as substrate, resulted in a peak of activity at mitosis in synchronised cells and essentially non-detectable activity in other phases [38], and a similar coincidence of activity with initiation of mitosis has been observed in *N. plumbaginifolia* (Fig. 3). These correlations indicate that p34^cdc2, perhaps supported by close variants, drives progress through mitosis in plants as in other eukaryotes (reviewed by P Nurse [54]) illustrated in Fig. 5. There is evidence that the peak of activity which is detected by *in vitro* assay, is associated with *in vivo* changes of phosphorylation because the MPM2 monoclonal antibody raised against phosphorylated nuclear proteins of mitotic HeLa cells only detects phosphorylated nuclear proteins in *Chlamydomonas* during mitotic phase [28].

Cyclin B-like proteins have been found to associate with purified plant p34^cdc2 and to be tightly coupled in their abundance during the cell cycle to the activity of p34^cdc2 both during normal mitosis and in mitotically blocked cdc mutant cells. The cyclins may direct the catalytic subunit p34^cdc2 to specific substrates, influence its phosphorylation and hence activity state and terminate its activity by being subject to proteolysis. Cyclins share a cyclical pattern of abundance during the cell cycle but at least three groups of cyclins must be recognised [60]. The cyclins of type CLN increase in abundance during G1-phase, affect the timing of START [61] and may play a part in directing p34^{cdc2} to the, currently unknown, substrates that are presumed to be phosphorylated at START of division (see D Dudits *et al.*, this volume). Cyclins of type A may direct activity to proteins necessary for DNA replication and may also have a negative role in preventing mitotic activation of p34^{cdc2} (which occurs by association with B cyclins) until replication is complete [67]. Cyclins of type B participate in the sudden activation of p34^{cdc2} that drives progress through prophase to metaphase of nuclear division. It remains unknown what mechanisms control the formation of different cyclin complexes with p34^{cdc2}. Present evidence of the timing of events at mitosis in animal and fission yeast cells indicates that p34^{cdc2} binds cyclin B in G2-phase, is then phosphorylated at *Tyr*15 and in higher eukaryotes also at *Thr*14, but is not activated with respect to mitotic substrates until these phosphates are removed. Activity of the p34^{cdc2}-cyclin B complex has a brief duration that is programmed by its phosphorylation of the cyclin B component, which is then degraded leaving monomeric p34^{cdc2} that is inactivated. Inactivation of p34^{cdc2} is necessary for completion of mitosis [52, 24]. While it is activated, p34^{cdc2} is a workhorse protein kinase that establishes metaphase by phosphorylating and so altering the binding properties or catalytic activities of several key proteins, including: (i) H1 histones and other chromosomal proteins that promote chromosome condensation; (ii) cytoskeletal proteins and kinases that may act on them, which promotes a transition from interphase microtubule arrays to establishment of a mitotic spindle; (iii) lamins, which promotes nuclear envelope breakdown (reviewed [54, 14]). The effect of okadaic acid indicates that interaction of kinase activity with phosphoprotein phosphatases is necessary for completion of the intranuclear events. We shall argue that in plant cells the mitotic disassembly of the preprophase band of microtubules, which is peculiar to plants, is also regulated by mitotic protein kinase activity.

Evidence from the plant kingdom can help to emphasise the relationship between B cyclin breakdown and the inactivation p34^{cdc2} that is necessary for progress from metaphase through anaphase. Although it is not known what molecular mechanism causes p34^{cdc2} enzyme activity to decrease sharply in anaphase a possible contribution from B cyclin breakdown is indicated by temporal coincidence [15] and by the ability of non-degradeable truncated cyclin in *Xenopus* egg extracts to cause persisting activity and arrest in a metaphase-like state [51]. In the plant mitotic cycle a tight coupling of cyclin degradation to p34^{cdc2} inactivation and exit from metaphase is indicated by a conditional cell division cycle (*cdc*) mutant of *Chlamydomonas*. Such *cdc* mutations, after repeated backcrossing to wild type, yield a cell in which a single protein is unusually thermolabile [59]. On transfer to an elevated temperature the effects of losing the function of a single *cdc* protein are seen. In the *met*1 mutant, cells arrest with a full mitotic spindle and with chromosomes aligned at the metaphase plate [38]. This mutation may be valuable because its precise equivalent has not yet been isolated in yeasts, although a similar block can be obtained in *Xenopus* egg extracts by adding undegradeable cyclin. We have compared *met*1 and wild type cells with respect to their level of cyclin B protein and activity of p34^{cdc2}-like protein kinase. Cyclin B was detected by cross reaction with antibody raised against p56^{cdc13}/cyclin B protein of *S. pombe*, which detected in *Chlamydomonas* a protein of identical 56 kDa size. Activity of p34^{cdc2}-like kinase was measured after affinity purification using p13^{suc1} binding and elution and was assayed in the absence of free calcium and cyclic AMP [36]. Both wild type and *met*1 cells at the permissive temperature of 21°C showed a cyclic presence of B cyclin, its abundance peaking at mitosis and coinciding with a peak of p34^{cdc2} H1 histone kinase. This temporal correlation therefore extends to the plant kingdom, and it can be seen to result from tight coupling between cyclin level and p34^{cdc2} activity because at the restrictive temperature proteolysis of the 56 kDa protein immunologically related to B cyclin was blocked and high and persisting p34^{cdc2}-like kinase activity was observed. In extracts from blocked cells the p34^{cdc2} kinase activity was more than six times higher than from synchronous culture at time of peak mitotic activity and this can probably be accounted for by the precise alignment of the population at metaphase, which can only be approximated in synchronous culture. The primary effect of the mutation is probably to prevent normal proteolysis of B cyclins. However we cannot conclude whether the machinery of breakdown is defective or whether there is a point in early metaphase prior to the cyclin degradation trigger at which the *met*1 mutant may be arrested. The effect of the mutation is specific to mitotic cyclins since cells can progress through all earlier phases of the cell cycle up to mitosis at the restrictive temperature. The mutation demonstrates in the plant kingdom the coupling of high B cyclin level with high activity of p34^{cdc2} kinase through prophase, as well as the necessity for cyclin degradation and p34^{cdc2} kinase inactivation for progress beyond metaphase.

Together with recently obtained sequence information indicating probable plant cyclin A and cyclin B genes [29] our evidence (and that of D Dudits *et al.*, this volume [13]) indicates that interactions between cyclins and p34^{cdc2}-like proteins are important in the plant cell cycle and the *met*1 mutant indicates that at mitosis the degradation of a B cyclin-like protein and inactivation of p34^{cdc2} H1 histone kinase is essential for anaphase.

5. Protein phosphorylation at mitosis: associated cytoskeletal rearrangements

Involvement of the p34^{cdc2} protein kinase enzyme in cell cycle control implies that phosphoprotein phosphatases will also be necessary to restore unphosphorylated proteins that can be responsive to further control. In animal and fungal cells protein phosphorylation is maximal at mitosis [39], which

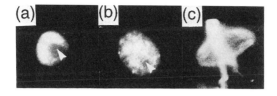

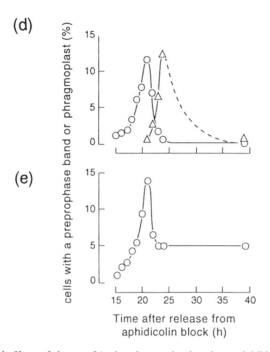

Fig. 4. Differential effects of the type 2A phosphoprotein phosphatase inhibitor okadaic acid on mitotic processes in suspension culture of *Nicotiana plumbaginifolia* previously synchronised by a 24 h aphidicolin block released at 0 h in the experimental sampling period. (a) Typical nuclear configuration of interphase cell, with Hoechst- stained DNA, showing absence of chromosome condensation and presence of nucleolus (arrowed). (b) Nucleus typical of cells treated with 12 μM okadaic acid from 0 h, all of which accumulated with partially condensed chromosomes and persisting nucleolus (arrowed) that is typical of early prophase. (c) Late prophase configuration of preprophase band (PPB) present with developing extranuclear spindle. This is the earliest phase of normal mitosis that was not attained in treated cells, which never developed a mitotic spindle. (d) Frequency of cells containing, ($\bigcirc$) a PPB, ($\triangle$) or phragmoplast, in the control population. (e) Incidence of PPB and phragmoplast (symbols as in (d) in cells treated with 12 μM okadaic acid from 0 h. Treated cells accumulated with an early prophase nuclear configuration and no mitotic spindle but 95% progressed to PPB disassembly without the usual accompanying spindle formation or establishment of metaphase, indicating differential sensitivity of nuclear and cytoplasmic processes that are dependent upon protein phosphorylation [71], see also text and Fig. 5.

indicates that dephosphorylation must occur in late mitosis, and in plants there is direct evidence for dephosphorylation of nuclear proteins at anaphase since the MPM2 monoclonal antibody, raised against nuclear phosphorylated proteins of mitotic HeLa cells, recognises equivalent phosphorylated proteins only in mitotic nuclei of *Chlamydomonas* [28]. Direct evidence for the importance of phosphoprotein phosphatases has come from the ability of mutations in genes with sequence homology to phosphatases of type 1 to block mitosis, as is the case with *bim*G in *Aspergillus* and *dis2/bws* and *sds*21 in fission yeast [11, 57, 5]. In particular the *cdc*25 gene of fission yeast, homologous with *string* of *Drosophila*, stimulates mitosis in a dose dependent manner [62], is necessary for dephosphorylation and activation of p34[cdc2] [24] and its product has been shown to have protein phosphatase activity [21]. Introduction of the gene to tobacco has pleiotropic effects [3].

To investigate the importance of phosphoprotein phosphatase activity in higher plant cells we have used okadaic acid to inhibit plant protein phosphatase type 2A [46] and found that it alters phosphoprotein levels when applied to whole cells of *Nicotiana plumbaginifolia* in suspension culture [71]. In inhibited cultures a decline in frequency of cells with G1 nuclear DNA content and accumulation of cells entirely with G2 content reflected a specific block to mitosis. Arrested cells had a nuclear configuration that resembled early prophase, with partially condensed chromosomes and persisting nucleolus, but never progressed to full chromosome condensation or formation of a mitotic spindle (Fig. 4). The arrest configuration was reached in synchronous culture at the same time that normal prophase was reached in control cells. The observation that disturbance of phosphoprotein levels specifically disrupts progress through prophase correlates with an observed peak in activity of p34[cdc2]-like kinase activity at that time, which provides a different form of evidence that protein phosphorylation is then particularly important. Direct assay of the p34[cdc2]-like protein obtained by p13[suc1] binding and elution showed that mitotic activation occurs at the same time in the okadaic acid treated cells (Fig. 3) but reached higher levels because the population became uniformly held in prophase. The capacity to activate p34[cdc2]-like enzyme activity in the presence of okadaic acid has a parallel in *Xenopus* egg extracts [19] and presumably indicates that the necessary *cdc*25-like phosphatase activity is insensitive to the inhibitor *in vivo*. Failure to progress beyond early prophase in the presence of okadaic acid suggests that an interaction of phosphorylations and dephosphorylations is involved in establishing metaphase. Possible candidates for reactions affected by disturbance of normal phosphatase activity, include the cascade of protein kinases *nim*1, *wee*1 and *mik*1 and counteracting phosphatase *cdc*25 that control phosphorylation of p34[cdc2] (reviewed [54, 18, 63] and the multi-step processes of chromosome condensation and cytoskeletal change (Fig. 5). Much remains to be learned since protein kinases and phosphatases are often themselves regulated by phosphorylation, so allowing complex interactions.

Okadaic acid provides evidence that the disassembly of the preprophase band (PPB) of microtubules, which normally follows initiation of a mitotic spindle and coincides with nuclear envelope breakdown at the initiation of metaphase

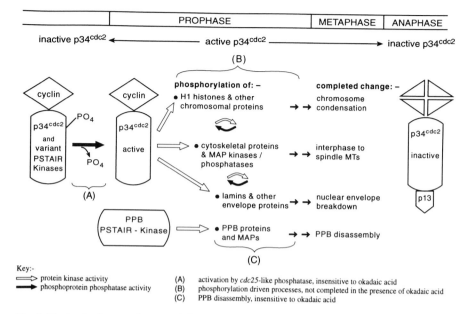

Fig. 5. Plant mitotic control; most of the core elements of a universal mitotic control mechanism (Nurse [54]), have now been detected in diverse plants, including the presence of *cdc2* genetic information [8, 17, 30, 31] and p34[cdc2] protein [37], its phosphorylation [37], affinity for p13[suc1] [36], association with cyclin (Fig. 2), transient activation of H1 histone kinase activity during mitosis (Fig. 3) tightly coupled to accumulation and breakdown of cyclin [38]. Involvement of p13[suc1]/CKS1 at anaphase is indirectly inferred from its affinity for plant p34[cdc2], its role in fission yeast and its evolutionary persistence in plants [36]. The processes of chromosome condensation, spindle formation, envelope breakdown and preprophase band (PPB) disassembly, although sequential do not operate as a dependent sequence in which completion of earlier events triggers each later event since differential sensitivity to okadaic acid prevents spindle formation but not PPB disassembly [71] (Fig. 4). It is inferred that the processes proceed independently following the common initiating stimulus of p34[cdc2] activation. Independence of prophase events is further indicated by microinjection experiments (Hepler P, Wu L, Dong C, Gunning B, John P, unpublished). The processes of chromosome condensation, spindle formation and nuclear envelope breakdown may be more sensitive to okadaic acid because they involve interactions between protein kinases and phosphatases. Disassembly of the PPB, which is peculiar to plants can be related to the observed p34[cdc2] kinase activation at mitosis. A PSTAIR protein that is presumably a member of the p34[cdc2]/CDK family has been detected in PPBs by two different antibodies ([48] and Fig. 6) but its activity at mitosis not demonstrated. See text and Figs. 2, 3, 4. Diagram from [38].

[25], does not depend upon the occurrence of these events since PPB disassembly can occur in cells that have not formed a mitotic spindle under okadaic acid inhibition (Fig. 4). The PPB is a key developmental determinant in plant cells since it marks the place where the future cross wall will join with the existing mother cell wall and therefore plays a part in determining the placing of new cells. The PPB must be cleared from the site when initiation of metaphase commits the formation of daughter nuclei that will be separated by new wall. Coordination of PPB disassembly with metaphase involves a positive

signal from an adjacent metaphase nucleus, which can be demonstrated by nuclear displacement experiments [53]. We have observed that PPB disassembly is coincident with activation of p34^{cdc2}-like protein kinase in tobacco (Figs. 3 and 4) and that in okadaic acid-inhibited cells continued activation of this enzyme correlates with uninhibited PPB disassembly. We suggest that the master control for progress through prophase is activation of p34^{cdc2} and related kinases and that a number of events then initiated normally occur in parallel but are not directly dependent on each other. A precedence for the independent progress of division events in plant division is provided by the continuation of cytokinetic events in *Chlamydomonas* when nuclear division is blocked [27]. The hypothesis that PPB disassembly can be triggered by activation of mitotic protein kinase independently of spindle formation is supported by the presence in the PPB of a PSTAIR containing protein, which is therefore a probable member of the p34^{cdc2}-kinase family [58] of mitotically activated protein kinases. The protein has been detected by a monoclonal antibody [48] and we have confirmed its presence (Fig. 6) (C Busby, B Gunning, P John, unpublished) using a different affinity-purified polyclonal antibody that has been verified to detect p13^{suc1}-affinity-purified plant p34^{cdc2}-like

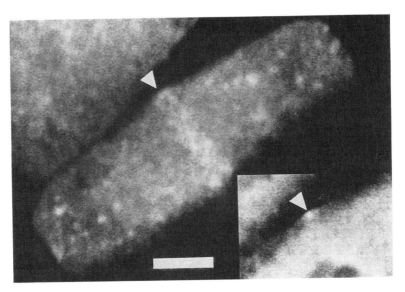

Fig. 6. Presence of PSTAIR-containing protein detected in the preprophase band (PPB) (arrowed) of onion root tips cells by indirect immunofluorescence using antibody raised against the peptide EGVPSTAIREISLLKE. Antibody was affinity purified by binding to immobilised authentic p34^{cdc2} and elution at low pH [37, 36]. This purified antibody detected p34^{cdc2} kinase in *Chlamydomonas*, wheat and tobacco ([37, 36] and Figs 2 and 7). Pre-competition of the antibody with 25 μM of PSTAIR peptide, which prevented reaction with p34^{cdc2} on Western blots [37, 36], prevented reaction with the PPB. The image was obtained by confocal microscopy as described [25]. Inset, plane of focus through centre of cell, which shows that the detected protein forms a ring in the cortical cytoplasm (arrowed). Scale bar 5 μm. Busby C, Gunning B, John P, unpublished.

protein kinase [36, 38] (Fig. 7). It is therefore likely that the protein kinase located in the PPB contributes directly to the phosphorylation signal for PPB disassembly. It is clear from microinjection of p34^{cdc2} protein kinase into animal cells [42] that this family of enzymes can cause disassembly of the interphase cytoskeleton. The differential effect of okadaic acid indicates that the protein kinase *in situ* may make a direct contribution to PPB phosphorylation without requiring the interactions with protein phosphatase activity that are necessary for establishing metaphase.

6. Changes in level of p34^{cdc2}-like protein and enzyme activity during differentiation and dedifferentiation

The growth of plants depends upon the restriction of cell division to specialised meristems, which produce cells with the frequency and orientation (PW Barlow, this volume [1]) that allows subsequent formation of specialised tissues and organs. Cells leaving meristem regions may not divide again but rather enlarge and take on specialised metabolic or structural functions. However division can be resumed in the formation of new meristems or in wound response. Therefore we investigated whether division proteins are retained during cell development and may be held inactive but available for resumption of division. Our studies of monocotyledonous and dicotyledonous leaves and roots have eliminated this possibility and shown that p34^{cdc2}-like proteins show more than fifteen-fold changes in level relative to other proteins during differentiation and dedifferentiation that are consistent with their operating as a developmental switch. Cells in which cell division is developmentally inappropriate establish low levels of p34^{cdc2}-like protein. Economy suggests that other division proteins will also be at low levels but the p34^{cdc2}-like proteins can be considered of primary importance for three reasons. Many proteins, like those encoded by *nim*1 and *wee*1, which participate in division control, have a fine tuning role by their effect on p34^{cdc2} but their absence does not prevent division [62]. The function of some division proteins can be substituted by enzymes of entirely unrelated sequence that are therefore not homologues of *cdc* genes and presumably not normally involved in the cell cycle, as human T cell protein tyrosine phosphatase can substitute for *cdc*25 [23]. However, p34^{cdc2} is essential at late G1 and also at mitosis when it is directly involved in many processes. Cells lacking p34^{cdc2} are therefore securely prevented from division and need not maintain other regulatory mechanisms for this purpose.

The developmental organisation that makes localisation of division important in plants also makes them experimentally more suitable subjects than are animals for analysing the molecular mechanism of transition from division to differentiation. Samples taken from the meristem and adjacent or older tissues provide a progressive series from division to differentiation. Seedling wheat leaves provide a linear developmental gradient from the meristematic

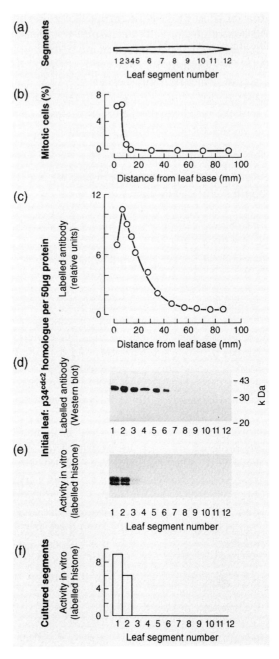

Fig. 7. Developmental progression and capacity for continuation of division by excised tissue in the wheat leaf, correlated with level of p34^{cdc2}-like protein and enzyme activity. (a) 90 mm long seedling leaf, showing the distribution of 4 mm segments taken for sampling. (b) Cell division activity indicated by percentage of cells in mitosis, detected by DNA staining with acetocarmine. (c) Level of p34^{cdc2}-like protein detected on Western blot fractionations of 50 μg samples of total

protein, using anti- PSTAIR antibody affinity purified against authentic p34[cdc2] and quantified by [125]I- labelled second antibody [35]. (d) Autoradiogram indicating distribution of p34[cdc2] that is presented in quantified form in (c) and shown here for visual comparison with distributions of enzyme activity. (e) Protein kinase activity of p34[cdc2]-like protein measured after affinity purification on p13[suc1]-agarose beads. Extract was pre- depleted with agarose beads, a volume of p13-beads giving linear recovery of p34 kinase was then used and, after two washes with 1% NP 40 detergent buffer then PBS, protein specifically bound to p13[suc1] was eluted with free p13[suc1] (essentially as described [36]) giving a considerable purification relative to the heterogenous protein fraction bound to the beads. The eluted kinase was assayed (as described [36] and in Fig. 3), using H1 histone (shown), or p34[cdc2]-preferred peptide ([47] and legend Fig. 2). Both were quantified after absorption onto P81 paper and phosphorylated histone was also measured after electrophoresis and analysis of the PhosphorImager image (shown). (f) Cultured leaf segments assayed for activity of p34[cdc2]-like protein kinase, measured as in (e). Leaf segments were transferred to MS agar [50] containing 0.6 μM 2,4-D and cultured for 14 days before assay. Activity was lower than in the meristem of intact plants because cell division was less frequent *in vitro*. At this low auxin analogue concentration only segments with high p34[cdc2]-like protein and activity at time of excision were able to continue p34[cdc2] accumulation in culture and form callus.

region at the base through progressive stages of differentiation to mature photosynthetic cells (Fig. 7). Mitotic activity is restricted to the lower 8 mm (Fig. 7 b) and coincides with maximum levels of extractable p34[cdc2]-like protein detected by PSTAIR antibody [35] (Fig. 7 c, d). Division also coincides precisely with the presence of p34[cdc2]-like enzyme activity (Fig. 7 e) measured after affinity purification by binding and elution with p13[suc1] (as in Fig. 3), again emphasising the importance of this enzyme in higher plant division. Mixing of tissue samples prior to p34[cdc2] extraction revealed no inhibitors of extraction, purification or assay in mature cells that could have caused an artifactual decline. The activity of the enzyme drops more sharply during cell differentiation than the level of its protein, implying that developmentally transitional cells cease dividing before reaching basal levels of the protein. Two mechanism probably contribute to this. The threshold level at which p34[cdc2] can support division activity is probably higher than the basal level, therefore when the threshold is reached cells cease to cycle and mitotically active p34[cdc2] enzyme is absent although residual protein may be detectable. Another probable contributary mechanism is that, in addition to the coarse control of division activity by extreme changes in p34[cdc2] protein level, cells that have recently been dividing retain fine control mechanisms (see for example earlier discussions of protein phosphorylation) that modulate division activity, perhaps in response to local phytohormone ratios.

The decline of p34[cdc2]-like protein in mature cereals cells is particularly significant since it becomes irreversible. When excised and transferred to MS agar medium [50], only cells that were in or adjacent to the meristem region of the leaf were able to respond to the presence of the auxin analogue 2,4-D by dividing and forming callus. A rapidly declining capacity for responding to the same medium is seen outside the meristem [68]. When transferred to medium with low 2,4-D concentration (0.6 μM) only cells containing active p34[cdc2] at time of excision could be stimulated to continue synthesis of p34[cdc2]-like

26

protein and form callus from which active enzyme was recovered (Fig. 7 f); all other tissue was unresponsive. At higher concentrations of 2,4-D (100 – 150 μM) cells immediately adjacent to the meristem that had recently ceased dividing and did not contain active p34^{cdc2} but did contain significant levels of the inactive enzyme protein could be stimulated to resume division (data not shown). The limits of response correlated with presence of p34^{cdc2} at above basal levels. A possible molecular mechanism can therefore be discerned for a differential cell division response to hormone by tissues from different regions [35].

The proposal of differential response by cells from different regions of the plant during a number of hormone-induced processes is of long standing [65] and has been accepted as one potentially important developmental mechanism [68, 20]. In the seedling wheat leaf differential division response could be accounted for by the simple mechanism that terminates p34^{cdc2} synthesis in

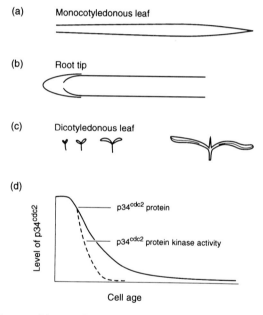

(a) Monocotyledonous leaf

(b) Root tip

(c) Dicotyledonous leaf

(d) Level of p34^{cdc2}

p34^{cdc2} protein

p34^{cdc2} protein kinase activity

Cell age

Fig. 8. Decline in level of p34^{cdc2}-like protein is a common pattern in plant development that is seen where a switch from division to cell differentiation involves extensive cell growth, as in: (a), cells leaving the basal meristem of monocotyledonous leaves ([35] and Fig. 7); (b), cells leaving the root tip meristem (Fig. 9, lanes 1-5); (c), cells terminating a phase of general cell division and initiating cell expansion and differentiation during organogenesis [22]. (d), the decline in p34^{cdc2}-like protein can often occur by cessation of p34^{cdc2} synthesis and dilution with other proteins. In the monocotyledonous leaf, decline in p34^{cdc2} activity and protein correlate with capacity to resume cell division at low or high auxin level in tissue culture (Fig. 7). The decline in p34^{cdc2} can be enforced, if dilution with other proteins does not achieve the decline as part of cell enlargement, as is seen when phytohormone stimuli signal termination of cell division in the root (Fig. 9). The decline in p34^{cdc2} is significant in maintaining the cessation of division since the enzyme has an indispensible role in the cell cycle (see text and Fig. 5) and its level must be restored prior to resumption of cell division (Figs. 9, 10).

cells that are displaced from the meristem followed by dilution of the protein. Measurement of the increase in total cell protein in the average cell during development shows that it is sufficiently extensive to account for the decline in p34^{cdc2} by dilution [35]. This simple mechanism of restricting the occurrence of high p34^{cdc2} level probably occurs widely in plant development. A close parallel occurs in the root tip, in which the relative level of p34^{cdc2} is high in the meristem region and declines as cells enlarge in the elongation zone (Fig. 9, lanes 1-5). We have also noted a similarly extensive increase in protein in the carrot cotyledon as the level of p34^{cdc2} declines and cell division terminates. Limitation of the distribution of high p34^{cdc2} levels by dilution can occur wherever the process of cell differentiation involves extensive increase in cell size. This common developmental motif is illustrated (Fig. 8). However the significance for division control of reducing p34^{cdc2} level relative to other

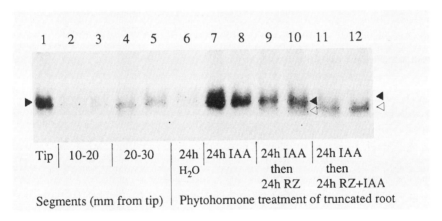

Fig. 9. Endogenous and induced levels of p34^{cdc2}-like protein in pea root tips and truncated roots. Primary roots of 50 mm length on seedlings of pea var. Alaska after 3 day germination in vermiculite were either segmented into regions 0-10 mm, 10-20 mm and 20-30 mm from the tip and analysed immediately (lanes 1-5), or alternatively roots were truncated by removing the tip 10 mm then, in the non-dividing elongating cells at the end of the truncated root, hormones and hormone mixtures known to affect induction of lateral root meristems [69] were tested for their effects on levels of p34^{cdc2}-like protein in the terminal 10 mm, previously 10-20 mm from the tip of the intact root. Samples containing 50 μg total protein were analysed for content of p34^{cdc2} on Western blots using affinity purified antibody as in Fig. 7 and described previously [35]. Abbreviations: IAA, 3-indolylacetic acid at 50 μM: RZ, Zeatin riboside at 50 μM. High levels of p34^{cdc2}-like protein (closed arrow head) were present in the tip meristem (lane 1) but this declined to a minimum in elongating cells 10-20 mm behind the tip (2&3) and began to reappear where secondary thickening and lateral root primordia begin to develop 20-30 mm behind the tip (4&5). Both p34^{cdc2}-like protein and lateral root primordia are induced by auxin within 24 h in the region that was at 10-20 mm from the original tip but was left at the end of the root after truncation (water control lane 6, IAA 7&8), but this can be reversed by subsequent treatment for 24 h with zeatin riboside alone (9&10) and particularly by the riboside in combination with auxin (11&12), leading to extensive loss of antibody-reacting protein and to a smaller size and greater electrophoretic mobility in the residual antibody-binding protein that is presumably a breakdown intermediate (open arrow head).

28

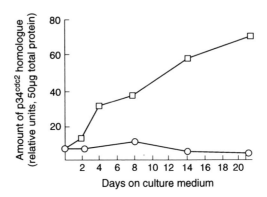

Fig. 10. Increase in level of p34^{cdc2}-like protein during induction of cell division in excised segments of mature cotyledon tissue from 11 day old carrot seedlings. p34^{cdc2}-like protein was quantified as in Fig. 7. Segments were cultured on MS agar containing; (□), 0.45 μM 2,4-D or; (○), no 2,4-D. The auxin analogue stimulated a progressive accumulation of p34^{cdc2} and a spread of cell division and callus formation radiating outward from the vascular strands. In the absence of added hormone cells enlarged and became highly vacuolate but did not divide. At the end of the *in vitro* culture period, when extensive callus formation had occurred, levels of p34^{cdc2}-like protein were 70% of those in cotyledons of intact plants at two days after germination when essentially all cells are dividing actively (Redrawn from Gorst [22]).

proteins is underlined by the necessity for restoring the level prior to resumption of division in cells from the elongation zone of root tips (Fig. 9) and from differentiated leaf tissue (Fig. 10).

Although the level of p34^{cdc2} may often be controlled simply by regulation of its synthesis coupled with continued cell growth, a latent capacity for its active disposal has been revealed by study of lateral primordium induction in pea roots. In the primary seedling root removal of the tip 10 mm leaves a truncated root in which lateral root primordia can be induced by auxin [69] with a accompanying more than eight-fold induction of p34^{cdc2}-like protein within 24 h. If truncated roots that have been induced in this way are treated with the natural cytokinin, zeatin riboside, there is a disappearance of three quarters of the p34^{cdc2}-like protein within 24 h and development of the primordia is halted (P John, F Wightman, unpublished). This disappearance of p34^{cdc2} cannot be explained by dilution since little growth occurs in the cytokinin treated root, rather breakdown of the protein must occur and this correlates with the detection of a lower molecular weight fragment that may be a degradation intermediate.

This interaction of cytokinin with auxin in determining p34^{cdc2} level in the pea root eliminates the possibility that auxin might be the key determinant of its level. In this tissue both hormones influence the level of p34^{cdc2} but it remains to be established whether one or other hormone has a greater effect on synthesis or breakdown of the protein. As with other hormone induced phenomena in plants it is not clear how directly either auxin or cytokinin affect the *cdc2* gene. Hormone receptor molecules may be directly targeted to the *cdc2* gene but changes in its expression could be early secondary effects of more immediate responses, which may be clarified when the nature of phytohormone receptors is established in plants.

In the localised regions of high p34^{cdc2} level to which cell division is restricted we consider that the cell cycle is making several important contributions to development. (i) In stabilising cell size, and so partitioning new cytoplasm into cellular units with physiologically effective ratios of DNA to cytoplasm, efficient biosynthetic metabolism is maintained. (ii) Size at division can be altered as cells are displaced towards the periphery of the meristem and initials are established that are committed to form particular cell types. Initials of different cell types can have different sizes in their final rounds of division, as has been quantified for the outer cortical and endodermal cells of *Azolla* root [26]. The size of the resulting daughters allows them to elongate as the surrounding tissue expands and to attain the shape that is appropriate for their differentiated function. We postulate that an early event in differentiation occurs while initials are still dividing and is the setting of a critical size threshold for a cell cycle control point which results in daughter cells of a size that is appropriate for completion of their differentiation program. (iii) Active division is also associated with local phytohormone synthesis and therefore the resulting hormone-stimulated nutrient import completes a positive feedback cycle promoting continuing growth and further division, which can account for the commonly observed persistence of meristems within the life of the plant [66]. (iv) Divisions are precisely located within the meristem and do not occur at random, as they may do in a callus. The molecular basis of the positional controls remains unknown (see also D Francis and RJ Herbert, this volume), although important progress is being made in analysing the distribution of division (PW Barlow, this volume [1]).

In most monocotyledonous tissues the decline in p34^{cdc2} levels outside the meristem has the additional significance of being irreversible even if cells are stimulated by adjacent wounding or the artificial application of hormone. This underlines their difference in developmental strategy from dicotyledonous plants which commonly employ secondary thickening, activation of lateral meristems and wound-induced callus formation.

In dicotyledonous plants, the capacity of cells to respond to exogenous hormone, when taken into culture from organs of normally determinate growth, provides an opportunity to study the termination of division and its resumption when stimulated by phytohormone. The carrot cotyledon completes most of its cell division in the first two days after germination and then enters a period of more than ten-fold cell enlargement and differentiation to photosynthetic activity. A more than fifteen-fold decline in p34^{cdc2}-like protein coincided with cessation of division [22], as in wheat. When segments of 11 day old cotyledons were transferred to nutrient agar containing 2,4-D an induction of p34^{cdc2}-like protein, beginning with no discernable lag, preceeded dedifferentiation and resumption of division that commenced in cells adjacent to the vascular tissue (Fig. 10). The spread of division in the excised segment correlated with steady increase of p34^{cdc2}-like protein and eventually lead to the formation of a callus. This induction of cell division is significant in two ways; it provides a positive correlation between increase of p34^{cdc2}-like protein and induction of division, and also shows that mature differentiated cells were potentially able to accumulate p34^{cdc2} and therefore when ceasing to do so in early organogenesis were operating a developmental switch. We further tested whether loss of p34^{cdc2}-like protein was a primary cause of ceasing division, or perhaps was an inevitable secondary consequence of division having been prevented by some other means. We used a suspension culture, which was derived from the callus induced on agar, in which we could impose cessation of division by limiting nitrogen or carbon source and, although protein per fresh weight was reduced to 23% and 57% respectively, the same level of p34^{cdc2}-like protein was retained even though the reduced protein per fresh weight indicated that dispensable proteins had been subjected to turnover. We conclude that the cells in suspension culture continued to be developmentally programmed to divide, perhaps due the presence of 2,4-D. Since decline in p34^{cdc2}-like protein is not an inevitable secondary consequence of ceasing division, therefore where the decline occurs it can be seen as a positive developmental control.

7. Conclusion

The plant cell division cycle can be influenced by cell size and this provides a means for coordinating rates of division with rates of growth and so stabilising cell size. The key regulatory control points of the plant cell cycle are in late G1-phase and at the initiation of mitosis. Changes in the amount, phosphorylation and activity of p34^{cdc2}-like protein at these times, and temporal and *in vitro* associations of p34^{cdc2} protein with two regulatory binding proteins immunologically related to p13^{suc1} and p56^{cdc13} seen during normal division and in conditionally blocked *cdc* mutants, indicate that the plant kingdom shares a p34^{cdc2}-based cell cycle. The central role of this protein kinase correlates with the importance of normal levels of protein phosphorylation for completion of mitosis. The selective inhibition of individual events in mitosis revealed that the normal sequence, of chromosome condensation, spindle formation, nuclear envelope breakdown and PPB disassembly, does not derive from the serial activation of later events by the preceding ones. Rather these are independent processes that occur in parallel following an initiating stimulus that is probably the activation of p34^{cdc2} that we observe in prophase.

In higher plant development the presence of p34^{cdc2}-like protein is restricted to meristem tissues and to stages of development in which division is appropriate. Reduction of p34^{cdc2} level can often occur by dilution due to cell enlargement in differentiation but it can be imposed by breakdown of the protein if growth is more limited. There is a consequent exit from the cell cycle because of the essential role of the division protein in late G1 and at mitosis, and in normal plant development decline of the key cell cycle protein establishes a secure exit from the cell cycle that is necessary for differentiation. Resumption of division follows restoration of p34^{cdc2} levels and both auxin and cytokinin phytohormones can influence induction of the protein.

References

1. Barlow PW (1992) Supracellular controls of the cell division cycle in root development. Plant Growth Regulation, (in press).
2. Beach D, Durkacz B and Nurse P (1982) Functionally homologous cell cycle control genes in budding and fission yeast. Nature 300: 706-709.
3. Bell MH, Halford NG, Francis D and Ormrod JC (1992) Tobacco plants (in press) transformed with cdc25, a mitotic inducer from fission yeast. J Exp Bot.
4. Berger JD (1989) The cell cycle in lower eukaryotes. Current Opinion Cell Biol 1: 256-262.
5. Booher R and Beach D (1989) Involvement of a type 1 protein phosphatase encoded by $bws1^+$ in fission yeast mitotic control. Cell 57: 1009-1016.
6. Brizuela L, Draetta G and Beach D (1987) p13^{suc1} acts in the fission yeast cell cycle as a component of the p34^{cdc2} protein kinase. EMBO J 6: 3507-3514.
7. Broek D, Bartlett R, Crawford K and Nurse P (1991) Involvement of p34^{cdc2} in establishing the dependency of S phase on mitosis. Nature 349: 388-393.
8. Colsanti J, Tyers M and Sundaresan V (1991) Isolation and characterisation of cDNA clones encoding a functional p34^{cdc2} homologue from *Zea mays*. Proc Natl Acad Sci USA 88: 3377-3381.
9. Donnan L, Carvill EP, Gilliland TJ and John PCL (1985) The cell cycles of *Chlamydomonas* and *Chlorella*. New Phytol 99: 1-40.
10. Donnan L and John PCL (1983) Cell cycle control by timer and sizer in *Chlamydomonas*. Nature 304: 630-633.
11. Doonan JH and Morris NR (1989) The *bim* G gene of *A. nidulans*, which is required for completion of anaphase, encodes a homolog of mammalian protein phosphatase 1. Cell 57: 987-996.
12. Draetta G and Beach D (1988) Activation of cdc2 protein kinase during mitosis in human cells: cell cycle dependent phosphorylation and subunit rearrangement. Cell 54: 17-26.
13. Dudits D, Bako L, Magyar Z, Bogre F, Felfoldi F, Deak M, Dedeoglu D and Georgyey J (1992) Key components of cell cycle control during auxin induced cell division. Plant Growth Regulation, (in press).
14. Enoch T, Matthias P, Nurse P and Nigg EA (1991) p34^{cdc2} acts as a lamin kinase in fission yeast. J Cell Biol 112: 797-807.
15. Evans T, Rosenthal ET, Youngblom J, Distel D and Hunt T (1983) Cyclin: A protein specified by material mRNA in sea urchin eggs that is destroyed at each cleavage division. Cell 33: 389-396.
16. Fantes PA and Nurse P (1981) Division timing: controls models and mechanisms. In: The Cell Cycle. John PCL (ed), pp. 11-33. Cambridge, Cambridge University Press.
17. Feiler HS and Jacobs TW (1990) Cell division in a higher plants: a cdc2 gene, its 34 kDa product and histone H1 kinase activity in pea. Pro. Natl Acad Sci USA 87: 5397-5401.
18. Feilotter H, Nurse P and Young PG (1991). Genetic and molecular analysis of cdr1/nim1 in *Schizosaccharomyces pombe*. Genetics 127: 309-318.
19. Felix M-A, Cohen P and Karsenti E (1990) Cdc2 H1 kinase is negatively regulated by a type 2A phosphatase in the *Xenopus* early embryonic cell cycle: evidence from the effects of okadaic acid. EMBO J 9:675-683.
20. Finkelstein R, Estelle M, Martinez-Zapater J and Sommerville C (1988) *Arabidopsis* as a tool for the identification of genes involved in plant development. In: Temporal and Spatial Regulation of Plant Genes. Verma DPS and Goldberg RB (eds) pp. 1-25. Springer, New York.

21. Gautier J, Solomon MJ, Booher RN, Fernando Bazan J and Kirschner MW (1991) cdc25 is a specific tyrosine phosphatase that directly activates p34[cdc2]. Cell 67: 197-211.
22. Gorst J, Sek FJ and John PCL (1991) Levels of p34[cdc2]-like protein in dividing, differentiating and dedifferentiating cells of carrot. Planta 185: 304-310.
23. Gould KL, Moreno S, Tonks NK and Nurse P (1990) Complementation of the mitotic initiator p80[cdc25] by a human protein tyrosine phosphatase. Science 250: 1573-1576.
24. Gould KL and Nurse P (1989) Tyrosine phosphorylation of the fission yeast *cdc2*[+] protein kinase regulates entry into mitosis. Nature 342: 39-45.
25. Gunning BES (1992) Use of confocal microscopy to examine transitions between successive microtubule arrays in the plant cell division cycle. J Plant Cell Physiol, 7th Japan Prize Symposium. Shibaka H (ed), pp 145-155.
26. Gunning BES, Hughes JE and Hardham AR (1978) Formative and proliferative cell divisions, cell differentiation and developmental changes in the meristem of *Azolla* roots. Planta 143: 121-144.
27. Harper JDI and John PCL (1986) Coordination of division events in the *Chlamydomonas* cell cycle. Protoplasma 131: 118-130.
28. Harper JDI, Rao PN and John PCL (1990) The mitosis-specific monoclonal antibody MPM-2 recognises phosphoproteins associated with the nuclear envelope in *Chlamydomonas reinhardtii* cells. Eur J Cell Biol 51: 272-278.
29. Hata S, Kouchi H, Suzuka I and Ishii T (1991) Isolation and characterization of cDNA clones for plant cyclins. EMBO J 10: 2681-2688.
30. Hirayama T, Imajuku Y, Anai T, Matsui M and Oka A (1991) Identification of two cell-cycle controlling *cdc2* gene homologues in *Arabidopsis thaliana* Gene 109: 159-165.
31. Hirt H, Pay A, Georgyey J, Bako L, Newmeth K, Borge L, Schweyen RJ, Heberle-Bors E and Dudits D (1991) Complementation of a yeast cell cycle mutant by an alfalfa cDNA encoding a protein kinase homologous to p34[cdc2]. Proc Natl Acad Sci USA 88: 1636-1640.
32. Hori H, Lim B-L and Osawa S (1985) Evolution of green plants as deduced from 5S RNA sequences. Proc Natl Acad Sci USA 82: 820-823.
33. John PCL (1984) Control of the cell division cycle of *Chlamydomonas*. Microbiol Sci 1: 96-101.
34. John PCL (1987) Control points in the *Chlamydomonas* cell cycle. In: Algal Development (Molecular and Cellular Aspects). Wiessner W, Robinson DG and Starr RC (eds), pp. 9-16. Berlin, Springer-Verlag.
35. John PCL, Sek FJ, Carmichael JP and McCurdy DW (1990) p34[cdc2] homologue level, cell division, phytohormone responsiveness and cell differentiation in wheat leaves. J Cell Sci 97: 627-630.
36. John PCL, Sek FJ and Hayles J (1991) Association of the plant p34[cdc2]-like protein with p13[suc1]: implications for control of cell division cycles in plants. Protoplasma 161: 70-74.
37. John PCL, Sek FJ and Lee MG (1989) A homologue of the cell cycle control protein p34[cdc2] participates in the cell division cycle of *Chlamydomonas* and a similar protein is detectable in higher plants and remote taxa. Plant Cell 1: 1185-1193.
38. John PCL and Wu L (1992) Cell cycle control proteins in division and development. Journal Cell Plant Physiol, 7th Japan Prize Symposium. Shibaka H (ed), pp 65-84.
39. Karsenti E, Bravo R and Kirschner MW (1987) Phosphorylation changes associated with early cycle in *Xenopus* eggs. Dev Biol 119: 423-453.
40. Korner Ch, Pelaez Menendez-Riedl S and John PCL (1989) Why are Bonzai plants small? A consideration of cell size. Aust J Plant Physiol 16: 443-448.
41. Krek W and Nigg EA (1991) Differential phosphorylation of vertebrate p34[cdc2] kinase at the G1/S and G2/M transitions of the cell cycle: identification of the major phosphorylation sites. EMBO J 10: 305-316.
42. Lamb NJC, Fernandez A, Watrin A, Labbe J-C and Cavadore J-C (1990) Microinjection of p34[cdc2] kinase induces marked changes in cell shape, cytoskeletal organisation and chromatin structure in mammalian fibroblasts. Cell 60: 151-165.
43. Lee MG, Norbury CJ, Spurr NK and Nurse P (1989) Regulated expression and phosphorylation

of a possible mammalian cell cycle control protein. Nature 333: 676-679.

44. Lee MG and Nurse P (1987) Complementation used to clone a homologue of the fission yeast cell control gene *cdc2*. Nature 327: 31-35.

45. Lorincz A and Carter BLA (1979) Control of cell size at bud initiation in *Saccharomyces cerevisiae*. J Gen Microbiol 113: 287-295.

46. MacKintosh C and Cohen P (1989) Identification of high levels of type 1 and type 2A protein phosphatases in higher plants. Biochem J 262: 335-339.

47. Marshak DR, Vandenberg MT, Bae YS and Yu IJ (1991) Characterisation of synthetic peptide substrates for p34[cdc2] protein kinase. J Cellular Biochem 45: 391-400.

48. Mineyuki Y, Yamashita M and Nagahama Y (1991) p34[cdc2] kinase homologue in the preprophase band. Protoplasma 162: 182-186.

49. Mitchison JM (1971) The Biology of the Cell Cycle, Cambridge, Cambridge University Press.

50. Murashige T and Skoog F (1962) A revised medium for rapid growth and bioassays with tobacco tissue cultures. Physiol Plant 15: 473-497.

51. Murray AW and Kirschner MW (1989) Cyclin synthesis drives the early embryonic cell cycle. Nature 339: 275-280.

52. Moreno S, Hayles J and Nurse P (1989) Regulation of p34[cdc2] protein kinase during mitosis. Cell 58: 361-372.

53. Murata T and Wada M (1992) Cell cycle-specific disruption of the preprophase band of microtubules in fern protonema: effects of displacement of the endosperm by centrifugation. J Cell Sci 101: 93-98.

54 Nurse P (1990) Universal control mechanism regulating onset of M-phase. Nature 344: 503-508.

55. Nurse P and Bissett Y (1981) Gene required in G1 for commitment to cell cycle and in G2 for control of mitosis in fission yeast. Nature 292: 558-560.

56. Nurse P and Fantes P (1981) Cell cycle controls in fission yeast: a genetic analysis. In: The Cell Cycle. John PCL (ed), pp. 85-98. Cambridge: Cambridge University Press.

57. Ohkura H, Kinoshita N, Migatani S, Toda T and Yanagida M (1989) The fission yeast *dis*[+] gene identified by a cold-sensitive mutation defective in chromosome disjoining encodes on of two putative type 1 protein phosphatases. Cell 57: 997-1007.

58. Pines J and Hunter T (1991) Cyclin dependent kinases: a new cell cycle motif? Trends Cell Biol 1: 117-121.

59. Pringle JR and Hartwell LH (1981) The *Saccharomyces cerevisiae* cell cycle. In: The Molecular Biology of the Yeast Saccharomyces. Strathern J, Jones E, Broach J (eds), pp. 97-142. New York, Cold Spring Harbor Laboratory.

60. Reed SI (1991) G1 specific cyclins: in search of an S phase promoting factor. Trends Genet 7: 95-99.

61. Richardson HE, Wittenburg C, Cross F and Reed SI (1989) An essential G1 function for cyclin-like proteins in yeast. Cell 59: 1127-1133.

62. Russell P and Nurse P (1987) The mitotic inducer *nim*1[+] functions in a regulatory network of protein kinase homologs controlling the initiation of mitosis. Cell 49: 569-576.

63. Sadhu K, Reed SI, Richardson HE and Russell P (1990) Human homolog of fission yeast cdc25 mitotic inducer is predominantly expressed in G2. Proc Natl Acad Sci USA 87: 5139-5143.

64. Taylor FJR (1987) Problems in the development of an explicit hypothetical phylogeny of the lower eukaryotes. Biosystems 10: 67-89.

65. Trewavas A (1981) How do plant growth substances work? Plant Cell Envir 4: 203-228.

66. Walbot V (1985) On the life strategies of plants and animals. Trends Genet 1: 165-169.

67. Walker DH and Maller JL (1991) Role for cyclin A in the dependence of mitosis on completion of DNA replication. Nature 354: 314-317.

68. Wernicke W and Milkovits L (1987) Rates of uptake and metabolism of indole-3-acetic acid and 2,4-dichlorophenoxyacetic acid by cultured leaf segments at different stages of development in wheat. Physiol Plant 69: 23-28.

69. Wightman F, Schneider EA and Thimann KV (1980) Hormonal factors controlling the initiation and development of lateral roots II. effects of exogenous growth factors on lateral

root formation in pea roots. Physiologia Plant 49: 304-314.

70. Van 't Hof J (1974) Control of the cell cycle in higher plants. In: Cell Cycle Controls. Padilla GM, Cameron IL and Zimmerman A (eds), pp. 77-85, Academic Press, New York.

71. Zhang K, Tsukitani Y and John PCL (1992) Mitotic arrest in tobacco caused by the phosphoprotein phosphatase inhibitor okadaic acid. Plant Cell Physiol 33: 677-688.

3. Isolation of *Arabidopsis* homologues to yeast cell cycle genes

FELICITY Z. WATTS, NEIL J. BUTT, ANNA CLARKE,
PHILIP LAYFIELD, JESSE S. MACHUKA, JULIAN F. BURKE and
ANTHONY L. MOORE

Abstract

The aims of our laboratory are the isolation and characterisation of cell cycle genes from *Arabidopsis*, with a view to understanding their role in plant growth and development. To this end we have been testing different molecular strategies in order to isolate and identify *Arabidopsis* homologues to yeast (*Schizosaccharomyces pombe* and *Saccharomyces cerevisiae*) cell cycle genes. Using various techniques we have identified *Arabidopsis* DNA sequences with homology to ubiquitin carrier proteins, and to the *S. pombe* genes, *cdc25*, *suc1*, *cdc22* and *suc22*. In this paper we describe the characterisation of the *Arabidopsis UBC*a gene and the identification of a *suc1* protein in *Arabidopsis* and pea. Preliminary results concerning the analysis of expression of these genes are also reported.

1. Introduction

1.1. The cell cycle and cell cycle genes in S. pombe and S. cerevisiae

Analysis of the eukaryotic cell cycle has recently become the focus of a substantial amount of attention. Of particular prominence in these studies have been the two yeasts *S. pombe* and *S. cerevisiae* and the large number of cell cycle mutants in these organisms isolated by Nurse *et al.*, [37] and Hartwell [19]. These studies have identified two important cell cycle control points, one in G1 and the other in G2, as well as regulatory genes involved in controlling the traverse of cells through these control points (Fig. 1; see J Hayles and P Nurse, this volume).

The *S. pombe cdc2* gene has been identified as being required for passage through both the G1 and G2 control points [38]. This gene is conserved in other organisms, and is homologous to a *S. cerevisiae* gene, namely *CDC28* [31]. These genes encode protein kinases which are themselves regulated via phosphorylation [e.g. 47, 15, 42]. Dephosphorylation of *cdc2* is required for its G2 activity, but once cells have passed through G2, *cdc2* becomes phosphorylated again prior to entry into G1. Control of the levels of phosphorylated *cdc2* in G2 occurs via *cdc25* [44], which has recently been shown to be a protein phosphatase [11]. This gene also has a homologue in *S.*

35

J.C. Ormrod and D. Francis (eds.), Molecular and Cell Biology of the Plant Cell Cycle, 35–44.
© 1993 *Kluwer Academic Publishers. Printed in the Netherlands.*

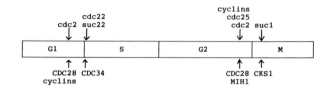

Fig. 1. Schematic diagram of genes required for cell cycle control.

cerevisiae, the *MIH*1 gene [45]. Another protein involved at this stage of the cell cycle is the *S. pombe suc*1 protein, the gene for which was isolated as a high copy number suppressor of the *cdc*2 mutation [21]. The *suc*1 protein forms a tight complex with *cdc*2 [3] and although its precise role in unknown, it is postulated to be required for mitotic kinase activity [33].

The *cdc*2 and *CDC*28 proteins complex with cyclins, which are specific for either G1 or G2 [1, 34, 35, 5, 16]. Most cyclins (exceptions include the mitotic cyclins in *S. pombe* [17]) show a cell cycle-dependent increase in concentration throughout the cell cycle peaking either in G1 or G2, after which time, the levels of the cyclins fall dramatically. Tagged sea urchin mitotic cyclin is ubiquitinated in *Xenopus* extracts prior to cyclin degradation [12], suggesting that ubiquitination may be required for controlling levels of cyclin proteins. Such ubiquitin-mediated proteolysis requires at least three different classes of genes encoding ubiquitin or polyubiquitin, ubiquitin conjugating proteins and ubiquitin ligases. It is thought that the ubiquitin conjugating proteins provide the specificity, i.e. target specific proteins for ubiquitination and hence proteolysis [26]. In *S. cerevisiae* more than eight ubiquitin carrier proteins and their corresponding genes have been identified [e.g. 26]. One of these, *CDC*34, at the non-permissive temperature, is defective in the G1-phase [13]. This therefore poses the question as to whether *CDC*34 has a role in ubiquitination of G1 cyclins.

The genetic regulation of the cell cycle occurs by a variety of mechanisms, e.g. at the levels of transcription and post-translational modification, such as phosphorylation (discussed above with regards to the control of *cdc*2 activity). Cell cycle regulated gene expression, most likely occurring at the level of transcription, has been shown for more than 27 genes in *S. cerevisiae*, including the G1 and mitotic cyclins [55, 49], DNA ligase (*CDC*9) [54], and the regulatory subunit of ribonucleotide reductase (*RNR*1) [8]. In contrast few cell cycle regulated genes have been identified in *S. pombe*, this is in keeping with the generally small number of inducible genes in this organism. This difference in number is important in relation to gene regulation in other organisms, such as higher plants. Four *S. pombe* genes which have been identified as being regulated during the cell cycle are: *cdc*22 – which encodes the regulatory subunit of ribonucleotide reductase [14], the genes which encode the histones, H2A and H2B [39] and *cdc25* [34]. These cell cycle-regulated genes are expressed only at specific points in the cell cycle, e.g. *cdc22* at G1/S, the histones genes (H2A and H2B) in S and *cdc25* at late S/G2. Deletion analysis and use of reporter gene fusions have been used to identify promoter elements responsible for this cell cycle control. In particular *Mlu*I elements promote expression at G1/S in *S. cerevisiae* when cloned upstream of the *E. coli lacZ* gene [32].

1.2. Conservation of cell cycle genes

Homologues to *cdc2*, *cdc25* and *suc1* have been identified in several higher eukaryotic organisms, e.g. *Drosophila, Xenopus* and mammals [e.g. 30, 7, 43]. In contrast to yeast where these genes are single copy, higher eukaryotes tend to have more than one copy of each gene, perhaps reflecting differences in function of the proteins in multicellular organisms. As in *S. cerevisiae*, mammalian cells contain several genes which are regulated during the cell cycle, and these include the cyclins [40, 41] and two *suc1* homologues [43].

1.3. Plant cell cycle genes

In contrast to extensive studies on plant development [e.g. 2], little is known about plant cell cycle genes. The exceptions are the identification of homologues to *cdc2* in various plants e.g. *Arabidopsis* [23, 10], pea [9] and wheat [27], cyclins in carrot, [20] and PCNA from soybean and rice [50]. The cDNAs were identified by a variety of methods e.g. hybridisation with yeast gene probes, PCR and antibody screening of plant cDNA expression libraries. Little information is available about whether cell cycle regulation of gene expression occurs in plants. However, it appears that in tomato and pea, the histone H2A gene is expressed in or near meristematic tissue, in a manner consistent with that of cell cycle regulation [28; see EY Tanimoto *et al.*, this volume].

In this paper we report on the identification of a gene encoding a ubiquitin carrier protein, attempts to identify *cdc25*, *suc1*, *cdc22* and *suc22* homologues, and preliminary analyses on the control of expression of these genes.

2. Materials and methods

2.1. Plasmids, libraries and strains

The *suc1*-containing plasmid was a gift from P Nurse (Oxford), *cdc22* and *suc22* containing plasmids were from P Fantes (Edinburgh). The lambda gt11 (cDNA) and EMBL3 (genomic) libraries were from Clontech and the lambda ZAP library was a generous gift of P Horsnell and C Raines (Essex). JE188 was from J Jones (Norwich). The *Arabidopsis thaliana* strain used was Columbia C24.

2.2. Isolation of Arabidopsis DNA and RNA

Genomic DNA was isolated using a modification of the method of Dellaporta *et al.* [6]. Prior to harvesting the plants were incubated for two days in the dark to minimise contamination of the preparation with chloroplast DNA and polysaccharides. Following grinding of tissue and incubation in extraction buffer, SDS was added to 1% followed by proteinase K to 10 mg/ml, and incubation was carried out at 65°C for 12 h. The solution was then subjected to centrifugation at 4 krpm for 10 min, the supernatant decanted to a fresh tube and DNA precipitated with 0.6 volumes of isopropanol. After redissolving in TE, the DNA was extracted three times with phenol/chloroform (50:50 mixture).

RNA was prepared according to the method of Hudson and modified as described by Holdsworth [24].

2.3. Molecular techniques

Standard molecular biology techniques were used, as described by Sambrook *et al.* [46]. Protein samples were prepared as described previously [53].

2.4. Antibody production, purification and Western analysis

*S. pombe suc*1 protein was purified from an *E. coli suc*1-overexpressing strain, by the method of Brizuela *et al.* [3]. Proteins or peptides conjugated to thyroglobulin were used to inject New Zealand white rabbits [18]. Purified antiserum was prepared by incubating with a nitrocellulose filter containing *suc*1 protein, non-specific IgG was washed off and then the purified anti-*suc*1 antibody was eluted. Western analysis was carried out following SDS-PAGE [29] and transfer of proteins to nitrocellulose [51]. Filters were incubated with antisera and stained with DAB, after treatment with GAR-PO.

2.5. Production of transgenic plants via Agrobacterium-mediated transformation of Arabidopsis roots

This was carried out using the method of Valvekens *et al.* [52].

3. Results

3.1. Strategies for the isolation of Arabidopsis genes

In order to efficiently isolate *Arabidopsis* homologues to yeast cell cycle genes, we have tested a number of different techniques. Potential methods included complementation of yeast cell cycle mutants, PCR, antibody screening of cDNA expression libraries and hybridisation with yeast genes. Any method used had to take into account that RNAs corresponding to cell cycle genes may be present in low amounts in dividing tissues but almost non-existent in expanding or mature cells. It was therefore necessary to take care when choosing the source of the material e.g. callus or very young seedlings which would be likely to consist of a higher proportion of dividing cells than mature plants. Additionally, care had to be taken not to amplify the libraries more than necessary since this may also lead to a loss of rare cDNAs.

Complementation of yeast mutations has been used to isolate a human homologue to *S. pombe cdc2* [30] and, earlier, for the isolation of a *Drosophila* adenine biosynthetic gene by complementation of the *S. cerevisiae ade*8 mutation [22]. A limitation to this technique is that *Arabidopsis* cDNA libraries must be constructed in a yeast vector. As these have only recently become available to carry out complementation experiments, this technique will not be discussed further here. The advantage of this strategy is that it will identify genes with specific functions rather than conserved domains or motifs.

The polymerase chain reaction (PCR) offers an excellent way of isolating genes conserved between organisms. It has been used to great effect in the isolation of plant cyclins [20] and a *cdc*2 homologue from *Arabidopsis* [10]. We have attempted to use the procedure to isolate sequences homologous to the *S. pombe cdc25* gene, as described below.

Antibody screening of plant expression libraries has been successful in the identification of a pea *cdc*2 homologue [9]. With this in mind, we have recently prepared antisera against a range of different yeast proteins (e.g. *S. pombe suc*1, myosin heavy chain protein, *rad*4 and *rad*9). We have been using these antisera for Western analysis of plant proteins and for screening plant cDNA expression libraries for the respective genes. Experiments with anti-*suc*1 antibodies have been particularly successful and are described below.

Hybridisation with known yeast genes is another technique which has been used to isolate *Arabidopsis* genes [36]. We have systematically screened Southern blots of *Arabidopsis* genomic DNA with known *S. cerevisiae* and *S. pombe* cell cycle and DNA repair genes with the result that approximately 50% of the genes we have tested show hybridisation to *Arabidopsis* DNA.

3.2. Identification of G2-related genes

In order to isolate an *Arabidopsis* homologue to *S. pombe cdc25*, mixed oligonucleotides corresponding to regions conserved between *S. pombe, S. cerevisiae* and *Drosophila* (as shown in Fig. 2) were synthesised and used for amplification of fragments from *Arabidopsis* genomic DNA. PCR fragments were cloned into M13mp18 and sequenced; the results from two fragments are shown in Fig. 2. These and other fragments are currently being used to screen *Arabidopsis* libraries to identify potential full-length clones.

The anti-*suc*1 antibodies have been used for Western analysis of proteins from *Arabidopsis* and pea. Figure 3 shows the result obtained using crude extract, a chloroplast/nuclei fraction, mitochondria and cytosol. In this case

```
S. pombe   I I D C R F E Y E Y L G G H I S T A V N L N T K Q A I V D A F - - - -
D. mel     . . . . . Y F . . . E . . . . E G A K . . Y . T E Q . L . E . L T V Q
S. cerev   . . . . . . . . . . . T . . . . I N S V . I H S R D E L E Y E . - - I H

At1                    Y E F G G G H I SYT L G R X N T S N R K H L Q R
At2        I I D C R - - Y N D L N V K M K T - - - L N T K V R Y L E G H


S. pombe   - - - - - - - - L S K P L T H R V A L V F H C E H S A H R A P H L A L
D. mel     Q T E L Q Q Q Q N A E S G H K . N I I I . . . . F . S E . G . K N S R
S. cerev   K V - L H S D T S N N N T L - P T L L I I . . . F . S H . G . S . . S

At1                    N R K Q L T(7)R G S A G S Q G T H S A
At2                    L A S S S A P K Q T S Q L P - R Q S N S E S K R I C S


S. pombe   H F R N T D R R M N S H R Y P F L Y Y P E V Y I L H G G Y K S
D. mel     F L . . L . . E R . T N A . . A . H . . . I Y L . . N . . . E
S. cerev   . L . . C . . I I . Q D H . . K . F . . D I L . . D G . . . A

At1
At2        C Y H T A X W X S T S N - - - - - - - - - - - - - X G G T K T
```

Fig. 2. Comparison of sequences for *cdc*25 homologues from *S. pombe, Drosophila melanogaster* (D. mel) and *S. cerevisiae* (S. cerev). Sequence of PCR primers used to amplify an *Arabidopsis* homologue to *S. pombe cdc*25 are shown as arrows. Two of the *Arabidopsis* sequences obtained (At1 and At2) are shown below. A dot represents a match, and a dash represents a gap introduced to maximise the alignment.

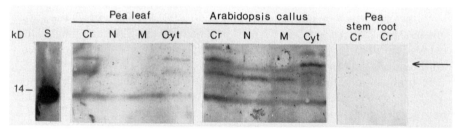

Fig. 3. Western analysis of proteins from *Arabidopsis* and pea. Proteins were separated on a 12% SDS polyacrylamide gel and blotted as described in materials and methods. The filters were incubated with anti-*suc*1 antibody (1/200 dilution) and processed as in materials and methods. S, *S. pombe suc*1 protein; Cr, crude extract; N, nuclei fraction; M, mitochondria; Cyt, cytosol; arrow indicates position of band observed using affinity purified anti-*suc*1 antibody.

the antibody has not been affinity purified and identifies a number of bands in the 13-18 kDa region. However use of affinity purified antibody (data not shown) identifies a protein in both *Arabidopsis* and pea at 18 kDa as indicated by the arrow. This is larger than that of *S. pombe suc*1 protein, however the size of *suc*1 protein appears to be quite different in different organisms, e.g. in mammals it is 9 kDa [43]. These anti-*suc*1 antibodies have been used to screen *Arabidopsis* cDNA expression libraries, and inserts from positive clones are currently being sequenced.

3.3. Analysis of G1 control sequences

In order to determine whether *Arabidopsis* contains genes which are regulated during the cell cycle and to determine which proteins interact with the promoters, we have initiated studies to isolate a homologue to *S. pombe suc22*, by hybridisation to the *S. pombe* gene. To date we have isolated positive clones and preliminary sequence data identifies a fragment with some homology to the yeast gene. We aim to use this to isolate the complete gene for analysis of the promoter. In parallel we intend to clone *Mlu*I elements [32] upstream of a minimal CaMV promoter (e.g. those in pBIN421.8 and pBIN421.9, M Bevan, J Innes, Centre for Plant Science Research, Norwich) to drive GUS expression in transformed callus and eventually, in transgenic plants. Localisation of GUS activity will be used to determine whether the *Mlu*I elements are sufficient to give cell cycle dependent expression.

3.4. Characterisation of members of the Arabidopsis UBC gene family

In view of its importance in regulating a number of cellular processes, we have recently isolated a member of the *Arabidopsis* ubiquitin conjugating protein gene family, *UBC*b (Fig. 4). DNA sequence analysis has been carried out, and the resulting major open reading frames identified. By comparison of the sequence of the predicted ORFs with those of known UBC proteins it appears that the gene comprises 6 exons and 5 introns. This is currently being confirmed through the identification of the corresponding cDNA. The putative splice sites are in good agreement with the plant consensus splice juctions [4]. Comparison of the sequence of the predicted protein with *S. cerevisiae UBC*8 indicates that *UBC*b has 35% homology in the N terminal region, between the *Nco*I sites (see Fig. 4) with 57% in the central region and almost no detectable homology in the C terminal part of the molecule. This gene is different to any of those identified in *Arabidopsis* by Vierstra *et al.* [48]. Additionally, this gene appears to be more similar to *UBC*8 than to *CDC*34 [13] or *RAD*6 (a gene required for DNA repair [25]). Southern analysis using this gene as a probe indicates that, as expected from results in other organisms, the gene is a member of a multi-gene family. Northern analysis shows that the gene is expressed at low levels and suggests that the protein may have a role in development. To further study the

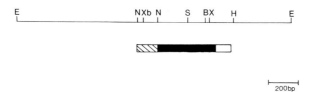

Fig. 4. Restriction map of *Arabidopsis UBC*b gene, showing percent homology to *S. cerevisiae* UBC8 gene. E, *Eco*RI; N, *Nco*I; Xb, *Xba*I; S, *Sal*I; B, *Bgl*II; X, *Xho*I; H, *Hind*III; hatched box indicates region with 35% identity to *S. cerevisiae* UBC8, filled box indicates region with 57% identity to UBC8 and open box indicates region with little detectable homology.

expression of the *UBC*b gene, promoter-GUS fusions have been constructed and used to produce transgenic plants. Preliminary studies appear to suggest that the gene is not expressed in mature leaves, however this analysis is not yet complete.

Acknowledgements

This work was funded by the AFRC. J. Machuka is supported by a British Council studentship. FZW and ALM would like to acknowledge receipt of Royal Society Travel Grants.

References

1. Booher RN, Alfa CE, Hyams JS and Beach DH (1989) The fission yeast cdc2/cdc13/suc1 protein kinase. Regulation of catalytic activity and nuclear localisation. Cell 58: 485-497.
2. Bowman JL, Smyth DA and Meyerowitz EM (1991) Genetic interactions among floral homeotic genes of *Arabidopsis*. Development 112: 1-20.
3. Brizuela L, Draetta G and Nurse P (1987) p13^{suc1} acts in the fission yeast cell division cycle as a component of the p34^{cdc2} protein kinase. EMBO J 6: 3507-3514.
4. Brown JWS (1986) A catalogue of splice junctions and putative branch point sequences from plant introns. Nucl Acids Res 14: 9549-9561.
5. Cross FR (1988) DAF1, a mutant gene affecting cell size control, pheromone arrest, and cell cycle kinetics of *Saccharomyces cerevisiae*. Mol Cell Biol 8: 4675-4684.
6. Dellaporta J (1992) Plant DNA extraction. In: *Arabidopsis*: the complete guide. D Flanders and C Dean (eds), Norwich: John Innes.
7. Edgar BA and O'Farrell PH (1989) Genetic control of cell division patterns in the *Drosophila* embryo. Cell 57: 177-187.
8. Elledge SJ and Davis RW (1990) 2 genes differentially regulated in the cell-cycle and by DNA damaging agents encode alternative regulatory subunits of ribonucleotide reductase. Genes and Dev 4: 740-751.
9. Feiler HS and Jacobs TW (1991) Cloning of the pea *cdc*2 homologue by efficient immunological screening of PCR products. Plant Mol Biol 17: 321-333.
10. Ferreria PCG, Hemerly AS, Villarroel R, Van Montagu M and Inze D (1991) The *Arabidopsis* functional homologue of the p34^{cdc2} protein kinase. Plant Cell 3: 531-540.
11. Gautier J, Solomon MJ, Booher RN, Bazan JF and Kirschner MW (1991) *cdc*25 is a specific tyrosine phosphatase that directly activates p34^{cdc2} Cell 67: 197-211.
12. Glotzer M, Murray AW and Kirschner MW (1992) Cyclin is degraded by the ubiquitin pathway. Nature 349: 132-138.
13. Goebl M, Yochem J, Jentsch S, McGrath JP, Varshavsky A and Byers B (1988) The yeast cell cycle gene CDC34 encodes a Ubiquitin-conjugating enzyme. Science 241: 1331-1335.
14. Gordon CB and Fantes PA (1986) the *cdc*22 gene of *Schizosaccharomyces pombe* encodes a cell cycle regulated transcript. EMBO J 5: 2981-2985.
15. Gould KL and Nurse P (1989) Tyrosine phosphorylation of the fission yeast *cdc*2^{+} protein kinase regulates entry into mitosis. Nature 342: 39-45.
16. Hadwiger JA, Wittenberg C, Richardson HE, de Barros Lopes M, Reed SI (1989) A family of cyclin homologs that control the G1 phase in yeast. Proc Natl Acad Sci USA 86: 6255-6259.
17. Hagan I, Hayles J and Nurse P (1988) Cloning and sequencing of the cyclin related *cdc*13 gene and a cytological study of its role in fission yeast mitosis. J Cell Sci 91: 587-595.

18. Harlow E and Lane D (1988) Antibodies: A laboratory manual. (Cold Spring Harbor, NY: Cold Spring Harbor Laboratory).

19. Hartwell LH (1974) Cell division from a genetic perspective. J Cell Biol 77: 627-637.

20. Hata S, Kouchi H, Suzuka I and Ishii T (1991) Isolation and characterisation of cDNA clones for plant cyclins. EMBO J 10: 2681-2688.

21. Hayles J, Beach DH, Durkacz B and Nurse P (1986) The fission yeast cell cycle control gene $cdc2^+$; isolation of a sequence $suc1^+$ that suppresses $cdc2^-$ mutant function. Mol Gen Genet 202: 291-293.

22. Henikoff S, Tatchell K, Hall BD and Nasmyth KA (1981) Isolation of a gene from *Drosophila* by complementation in yeast. Nature 289: 33-37.

23. Hirayama T, Imajuka Y, Anai T, Matsui M and Oka A (1991) Identification of two cell-cycle-controlling *cdc2* gene homologs in *Arabidopsis thaliana*. Gene 105: 159-165.

24. Holdsworth M (1991) *Arabidopsis* RNA extraction. In: *Arabidopsis*: the complete guide. D Flanders and C Dean (eds). Norwich: John Innes.

25. Jentsch S, McGrath JP and Varshavsky A (1987) The yeast DNA repair genes RAD6 encodes a ubiquitin-conjugating enzyme Nature 329: 131-134.

26. Jentsch S, Seufert W, Sommer T and Reins H-A (1990) Ubiquitin-conjugating enzymes: regulators of eukaryotic cells. Trends Biochem Sci 15: 195-198.

27. John PCL, Sek FJ, Carmichael JP and McCurdy DW (1990) p34[cdc2] homologue level, cell division, phytohormone responsiveness and cell differentiation in wheat leaves. J Cell Sci 97: 627-630.

28. Konig AJ, Tanimoto EY, Kiehne K, Rost T and Comai L (1991) Cell-specific expression of plant histone H2A genes. Plant Cell 3: 657-665.

29. Laemmli UK (1970) Cleavage of structural proteins during the assembly of the head of bacteriophage T4. Nature 227: 680-685.

30. Lee M and Nurse P (1987) Complementation used to clone a human homologue of the fission yeast cell cycle control gene *cdc2*. Nature 327: 31-35.

31. Lorincz AT and Reed SI (1984) Primary structure homology between the product of yeast cell division control gene CDC28 and vertebrate oncogenes. Nature 307: 183-185.

32. Lowndes NF, Johnson AL and Johnston LH (1992) Coordinated expression of DNA synthesis genes in budding yeast by a cell-cycle regulated *trans* factor. Nature 350: 247-250.

33. Moreno S, Hayles J and Nurse P (1989) Regulation of p34[cdc2] protein kinase during mitosis. Cell 58: 361-372.

34. Moreno S, Nurse P and Russell, P (1990) Regulation of mitosis by cyclic accumulation of p80[cdc25] mitotic inducer in fission yeast. Nature 344: 549-552.

35. Nash R, Tokiwa G, Annand S, Erikson K and Futcher B (1988) The WHI1[+] gene of *Saccharomyces cerevisiae* tethers cell division to cell size and is a cyclin homologue. EMBO J 7: 4335-4346.

36. Niyogi KK and Fink GR (1990) Two genes for anthranilate synthase in Arabidopsis. J Cell Biochem 15A: 78.

37. Nurse P, Thuriaux P and Nasmyth K (1976) Genetic control of the cell division cycle in the fission yeast *Schizosaccharomyces pombe*. Mol Gen Genet 146: 167-178.

38. Nurse P and Bissett Y (1981) Gene required in G1 for commitment to cell cycle and in G2 for control of mitosis in fission yeast. Nature 292: 558-560.

39 Osley MA (1991) The regulation of histone synthesis in the cell cycle. Ann Rev Biochem 60: 827-861.

40. Pines J and Hunter T (1989) Isolation of a human cyclin cDNA: evidence for cyclin mRNA and protein regulation in the cell cycle and for interaction with p34[cdc2]. Cell 58: 833-846.

41. Pines J and Hunter T (1990) Human cyclin A is adenovirus E1A-associated protein p60 and behaves differently from cyclin B. Nature 346: 760-763.

42. Reed SI, Hadwiger JA and Lorincz AT (1985) Protein kinase activity associated with the product of the yeast cell division cycle CDC28 gene. Proc Natl Acad Sci USA 82: 4055-4059.

43. Richardson HE, Stueland CS, Thomas J, Russell P and Reed SI (1990) Human cDNAs

44

encoding homologs of the small p34[cdc28/cdc2]-associated protein of *Saccharomyces cerevisiae* and *Schizosaccharomyces pombe*. Genes and Dev 4: 1332-1344.

44. Russell P and Nurse P (1987) Negative regulation of mitosis by *wee1+*, a gene encoding a protein kinase homolog. Cell 49: 559-567.

45. Russell P, Moreno S and Reed SI (1989) Conservation of mitotic controls in fission and budding yeast. Cell 57: 295-303.

46. Sambrook J, Fritsch EF and Maniatis T (1987) Molecular cloning: A laboratory manual, 2nd ed. Cold Spring Harbor, NY: Cold Spring Harbor Laboratory.

47. Simanis V and Nurse P (1986) The cell cycle control gene *cdc2+* of fission yeast encodes a protein kinase potentially regulated by phosphorylation. Cell 45: 261-268.

48. Sullivan ML and Vierstra RD (1991) Cloning of a 16-kDa ubiquitin carrier protein from wheat and *Arabidopsis thaliana*. J Biol Chem 266: 23878-23885.

49. Surana U, Robitsch H, Price C, Schuster T, Fitch I, Futcher B and Nasmyth K, (1991) The role of CDC28 and cyclins during mitosis in the budding yeast *S. cerevisiae*. Cell 65: 145-161.

50. Suzuka I, Hata S, Matsumoto M, Kosugi S and Hashimoto J (1991) Highly conserved structure of proliferating cell nuclear antigen (DNA polymerase δ auxiliary protein) gene in plants. Eur J Biochem 195: 571-575.

51. Towbin H, Staehelin T and Gordon J (1979) Electrophoretic transfer of proteins from polyacrylamide to nitrocellulose sheets: Procedure and some applications. Proc Natl Acad Sci 76: 313-340.

52. Valvekens D, Van Montagu M and Van Lijsebettens M (1988) *Agrobacterium tumefaciens*-mediated transformation of *Arabidopsis thaliana* root explants by using kanamycin selection. Proc Natl Acad Sci USA 85: 5536-5540.

53. Watts FZ, Walters AJ and Moore AL (1992) Characterisation of PHSP1, a cDNA encoding a mitochondrial HSP70 from *Pisum sativum*. Plant Mol Biol 18: 23-32.

54. White JHM, Barker DG, Nurse P and Johnston LH (1986) Periodic transcription as a means of regulating gene expression during the cell cycle: contrasting modes of expression of DNA ligase in budding and fission yeast. EMBO J 5: 1705-1709.

55. Wittenberg C, Sugimoto K and Reed SI (1990) G1-specific cyclins of *S. cerevisiae*: cell cycle periodicity, regulation by mating pheromone, and association with the p34[cdc28] protein. Cell 62: 225-237.

Note added in proof

Further analysis of clones At1 and At2 indicates that they do not encode parts of potential *Arabidopsis cdc25* homologues. In addition, recent communication with Vierstra and colleagues indicates that UBC6 is identical to one of their genes; our gene has therefore been numbered accordingly.

4. The molecular genetics and biochemistry of DNA replication

STEPHEN J. AVES and JOHN A. BRYANT

Abstract

In the budding yeast, *Saccharomyces cerevisiae*, genes encoding most of the recognised enzymes of DNA replication and nucleotide biosynthesis have been cloned. Molecular genetic analysis has also identified other genes whose functions are implicated in DNA replication. Reverse genetic studies have shown that three DNA polymerases (I, II & III – homologous to DNA polymerases α, ϵ & δ respectively of multicellular eukaryotes) are essential for viability. All *S. cerevisiae* genes encoding enzymes of the DNA replication complex are transcriptionally regulated and are co-ordinately expressed at G1/S. *Cis*- and *trans*-acting controlling factors have been identified which are conserved in the distantly related fission yeast *Schizosaccharomyces pombe*. Studies of plant DNA replication are much less advanced; although many enzyme activities have been detected only DNA polymerases and topoisomerases have been characterised. The existence of DNA polymerase α is well established in many plants. Although DNA polymerase δ has only been purified from wheat, its accessory protein, PCNA (proliferating cell nuclear antigen) has been identified in many plants and its gene cloned. A replicative complex has been isolated from pea shoot meristems which contains primase, ribonuclease-H, exonuclease, topoisomerase, protein kinase and DNA-binding activities. However, data are sparse concerning the control of activity of plant DNA replication enzymes at either the transcriptional or post-translational levels.

1. Introduction

The particular facet of the cell cycle with which we are concerned in this chapter is DNA replication. In a normal cell division cycle, DNA replication is confined to a specific phase of the cycle, the S-phase. In a typical somatic cell cycle, passage through S-phase results in the doubling of the DNA content from the 2C level to the 4C level (where 1C is the amount of DNA contained within the haploid genome). The histones are also synthesised in S-phase so that chromatin is re-assembled from the nascent DNA chains as replication progresses ([13,14], also see D Chiatante, this volume). Mitosis and cytokinesis (the M-phase of the cell cycle) are separated from DNA replication by G2, the

J.C. Ormrod and D. Francis (eds.), Molecular and Cell Biology of the Plant Cell Cycle, 45–56.

duration of each phase depending on plant species, growth conditions and location within the plant [27]. In addition to these 'normal' cell cycles there are many examples of the occurrence of DNA replication which is not followed by mitosis [52]. These truncated cycles, in which S-phase is uncoupled from cell division, lead to the endoreduplication of DNA. DNA endoreduplication, like cell division, is under specific developmental control, but in this instance the developmental control overrides the linkage between DNA replication and cell division which in 'normal' cells ensures that DNA replication occurs only once per cell cycle. The overall processes of plant development control the spatial and temporal distribution of DNA replication and endoreduplication and in turn are also likely to regulate, at a 'coarse' level, the activity of the biochemical machinery which synthesises DNA. Within an individual division or endoreduplication cycle there will also be more subtle levels of control, such as those which regulate enzyme activity within specific phases of the cycle.

To understand these various levels of control, it is necessary to investigate the genes which encode the proteins which are involved, and the proteins themselves. We start first with the genes, and here we will concentrate on the organisms which have proved the most useful for genetic studies of DNA replication, the yeasts.

2. Enzymes of DNA replication in yeast

Molecular genetic and biochemical studies on the enzymes involved in DNA replication are far advanced for the budding yeast *Saccharomyces cerevisiae*. Yeast homologues of most of the DNA replication or nucleotide biosynthesis

Table 1. DNA replication genes in *S. cerevisiae*

Gene	Gene product of function
*CDC*21	Thymidylate synthase
*CDC*9	DNA ligase
*CDC*8	Thymidylate kinase
*RNR*1	Ribonucleotide reductase, regulatory subunit
*POL*1	DNA polymerase I (α)
*PRI*1,2	DNA primase
*POL*2	DNA polymerase II (ϵ)
*DPB*2	DNA polymerase II, subunit B
*DPB*3	DNA polymerase II, subunit C
*POL*3 (*CDC*2)	DNA polymerase III (δ)
*POL*30	PCNA
*RFA*1,2,3	Replication factor A
*DBF*4	Interacts with *CDC*7 kinase

enzymes identified in animal systems have been purified, and many of the genes encoding these proteins have been cloned. Thus genes encoding three of the four subunits of the DNA polymerase I-DNA primase complex (DNA polymerase I is homologous to DNA polymerase α) have been cloned (POL1, PRI1 and PRI2 – see Table 1). All three genes are essential to the cell – the use of reverse genetics has shown that deletion of any one of them from the genome gives rise to a lethal phenotype [34, 45, 26].

Similarly, the DNA polymerase III (polymerase δ) catalytic gene POL3 and the gene POL30 encoding PCNA (proliferating cell nuclear antigen) have also been shown to be required for DNA replication [60, 10, 7]. Replication factor C has been purified from yeast cells [25], but the gene encoding this protein has not yet been cloned.

DNA polymerases α and δ are the only two polymerases required for replication of SV40 DNA in vitro [8], however the cloning of yeast genes has revealed that a third DNA polymerase is also necessary for DNA replication in vivo in this organism – DNA polymerase II, which is homologous to DNA polymerase ϵ of multicellular eukaryotes. This enzyme appears to be encoded by four genes, three of which have been cloned: POL2 and DPB2 are essential to the cell, while deletion of DPB3 is not lethal but leads to a higher spontaneous mutation rate, suggesting a role for this subunit in maintaining the fidelity of DNA replication [51, 2, 3]. The requirement for DNA polymerase II in yeast DNA replication has prompted models in which all three essential polymerases contribute to DNA synthesis at the replication fork [51, 4].

Molecular genetic studies are also identifying genes whose products are involved in the initiation of DNA replication in yeast. Replication factor A (RFA) is encoded by three genes in S. cerevisiae (RFA1, 2 and 3), all of which are essential for cell viability [11]. The products of several cell division cycle genes which act in late G1 may also be involved in the initiation of DNA replication – for example CDC7 (see M Buddles et al., this volume). The cell cycle gene DBF4 also falls within this category since evidence from genetic suppression and double mutant studies suggests that its product interacts with the CDC7 protein kinase [37]. In a similar way, interactions occur between the genes CDC45, 46, 47 and CDC54, mutants of which are defective in the initiation of DNA synthesis [31]. The CDC46 gene product is particularly interesting since its subcellular localisation is cell cycle regulated; in G1 it is predominantly nuclear but it migrates to the cytoplasm at the onset of S-phase, reappearing in the nucleus in late mitosis [30]. The CDC46 protein has extensive homology to the products of two further genes, MCM2 and MCM3, which were first identified as mutants defective in minichromosome maintenance [66, 47]. The MCM2 protein contains a zinc finger domain in a region essential for function, suggesting that it binds to DNA [66].

The availability in yeast of a number of ARS (autonomonously replicating sequence) elements which have been shown to act as origins of replication in vivo is a great advantage in the study of the initiation of DNA replication. Factors which specifically bind to such origins have been identified by gel retardation and nuclease protection studies. To date, however, such factors as *ABF*1 (= *OBF*1) have proved to be general transcription factors with a number of binding sites other than ARS elements [16].

3. Regulation of enzyme activity during the yeast cell cycle

Two lines of evidence suggest that post-translational regulation by protein phosphorylation is important in DNA replication. The first is the discovery that genes involved in DNA replication, such as *CDC*7, encode protein kinases [5]. The second is that a number of proteins involved in DNA replication are phosphorylated, and that phosphorylation status can change through the cell cycle. The second largest subunit of *RFA* is one such protein: it is phosphorylated at the G1/S-phase transition and dephosphorylated at mitosis, in both budding yeast and human cells [59]. Studies on phosphorylation in yeast DNA replication are at an early stage, however, and there is at present no accepted model for such control in S-phase.

The general application of immunolocalisation techniques to yeast cells is allowing the subcellular localisation of DNA replication factors to be studied through the cell cycle. It is of interest that the cell cycle-dependent localisation of the product of the *CDC*46 gene (see above) is consistent with the pattern expected of a factor whose presence could allow DNA replication, and whose absence prevents rereplication [9].

It is now clear that transcriptional control plays an important role in DNA replication in budding yeast. The mRNA levels of all cloned genes encoding enzymes of DNA replication, and several of the enzymes of nucleotide biosynthesis, show marked cell cycle fluctuations. At least 17 genes are regulated in this way, including all those in Table 1, with their expression occurring coordinately at G1/S (reviewed in ref. [35]). All of these genes have been found to possess copies of a hexanucleotide sequence ACGCGT (or closely related sequences) in their promoter regions. Because this sequence is an *Mlu*I site, these have been termed *MCB* elements (*Mlu*I Cell cycle Box). Deletion of these *MCB* elements from a gene abolishes G1/S-regulation of that gene, whereas addition of *MCB* elements confers cell cycle regulation on reporter genes, thus demonstrating that these are true cell cycle boxes [50, 42, 29].

A transcription factor has been detected which binds to *MCB*s. This transcription factor has been termed *DSC*1 (for DNA Synthesis Control), and *DSC*1 binding is cell cycle-dependent in *S. cerevisiae* [42]. One component of *DSC*1 has been identified as the product of the previously identified *SWI*6 gene; no binding activity is detected in *swi*6 deletion mutants [41]. *SWI*6 is a multifunctional protein which has been implicated in the control of transcription of other types of cell cycle genes. For example a complex of *SWI*6 and *SWI*4 proteins forms a transcription factor (*CCBF*) which controls expression of the G1 cyclin genes *CLN*1 and *CLN*2, and of the *HO* gene which controls mating-type switching [43]. The binding site for *SWI6/SWI4* is different from that of *DSC*1, so it is thought that *DSC*1 is composed of *SW*16 plus as yet unidentified factor(s) which confer *MCB*-binding specificity.

The *DSC*1 system appears to be conserved in the distantly-related fission yeast *Schizosaccharomyces pombe*. A *DSC*1-like factor which binds *MCB*s is present in this yeast [44]. The component of this transcription factor homologous to *SWI*6 appears to be the product of the fission yeast *cdc*10 START gene [44].

Despite the apparent conservation of a DNA synthesis control mechanism between distantly-related eukaryotes, there are important differences between the two yeast species. In particular, only one cell cycle-regulated DNA synthesis gene (*cdc*22, which encodes the large subunit of ribonucleotide reductase) has so far been detected in *S. pombe* (excluding the histone genes, which are regulated independently of the *DSC*1 control in both species [54]), and some enzymes such as DNA ligase are controlled at the transcriptional level in *S. cerevisiae* but not in *S. pombe* [65]. It therefore remains to be seen if a *DSC*1-type control exists in plant (or animal) cells, or whether it is an adaptation to rapid cell cycling in unicellular fungi.

Although many data are being collected on the regulation of DNA replication enzyme activity in the yeast model systems, it should be noted that most of the data pertain to cycling cells – very little information has been obtained in yeast on control of DNA synthesis on entry into, or exit from, the cell cycle.

4. Correlative studies between biochemistry and genetics

The previous sections of this chapter have illustrated clearly the usefulness of unicellular eukaryotes, and in particular the yeasts, in studying the molecular genetics of DNA replication. Genes have been identified by analysis of cell division cycle mutants and then isolated by mutant complementation. Analysis of the isolated genes has in several instances indicated the nature of the protein encoded and of the regulatory sequences involved in controlling gene expression. From this genetic approach it is now clear that DNA replication involves not only an array of enzymes directly involved in replication (see below) but also a range of proteins with a potential for regulating enzyme activity, and proteins involved in the interaction of the replicative enzymes with the DNA template, such as DNA-binding proteins. Further, in *S. cerevisiae*, there has also been extensive biochemical work which has led to the isolation, purification and characterisation of many of the gene products. In yeasts therefore, we are beginning to understand some aspects of DNA replication at all levels from control of the expression of the genes to the specific functions of enzymes in the replicative process.

5. Enzymology of DNA replications in plants

All the enzymes known to be involved directly in DNA replication (Table 2) have been detected in plants, but with the exception of DNA polymerase [13-15] (and to some extent topoisomerase [19]) there has been very little detailed characterisation of the plant enzymes. Five different DNA polymerases have been described in animal cells, (namely polymerases α, β, γ, δ and ϵ, the last-named being a very recently discovered class of polymerase [6]). For many years, polymerase-α was thought to be the only polymerase involved in replication of nuclear DNA, but it is known that polymerase-δ also has a major role [24]. DNA polymerase δ, in the presence of its auxiliary protein, PCNA is highly processive and in current models for animal cell DNA replication is thought to be part of a complex which synthesises the leading strand [23, 8]. DNA polymerase-α is closely associated with primase, and is not so processive. These features make polymerase-α highly suitable for synthesis of the lagging strand where it is envisaged to form part of a complex which is clearly separable from the polymerase-δ leading strand complex [8, 23]. As described in an earlier section, polymerase-δ and ϵ-like enzymes have also been discovered in yeast (and their genes have been isolated).

Table 2. Enzymes/proteins directly involved in DNA replication which have been detected in plants

Enzyme/protein	Comments
DNA-binding protein	Specificity not yet determined
Helicase	Promotes strand separation
Topoisomerase I	Presumed to relieve superhelical tension in front of the replication fork
Topoisomerase II	Possibly involved in resolution of replicated DNA molecules
DNA primase	Associated with DNA polymerase-α
DNA polymerase-α	The most extensively characterised plant DNA replication enzyme. Activity strongly correlated with DNA replication
DNA polymerase-δ	So far, only one unequivocal demonstration of the presence of this enzyme
PCNA	Transcription of the gene shows a marked peak in S-phase
$3' \rightarrow 5'$ exonuclease	Presumed to be a proof-reading activity
Ribonuclease-H	Removes RNA primers from nascent DNA molecules
DNA ligase	An unstable enzyme in plants proving very difficult to purify

The situation in plants is slowly being clarified. The existence of polymerase-α is certainly well established [13, 14] but, because very few investigators have attempted to assay specifically a polymerase-δ-like-enzyme, it is quite possible that many of the polymerase-α preparations also contained polymerase-δ. DNA polymerase-δ has so far been purified from only one plant, wheat [57]; like its counterpart in animals and like DNA polymerase III in yeast, it is highly processive and possesses 3′→5′ exonuclease (proof reading) activity. Several plants are known to contain the polymerase-δ auxiliary protein, PCNA [62, 63, 38], which has been identified at both the gene and protein level. Indeed, to date, the PCNA gene is the only gene encoding a protein directly involved in DNA replication which has been isolated from plants.

Replicative DNA polymerases in animals and yeast are part of large multi-protein complexes [33, 49]. A similar complex has been isolated from pea shoot meristems [15]. It has been described as a polymerase-α complex, but as the authors point out, it may be a mixture of the lagging strand polymerase-α complex and the leading strand polymerase-δ complex. The presence in the complex of primase, ribonuclease-H, exonuclease and topoisomerase is certainly consistent with this idea, but in the absence of a specific identification of polymerase-δ and/or PCNA, no firmer classification can be made.

A key issue in relation to the regulation of DNA replication, especially in view of the apparent flexibility in replicon spacing [12] is the interaction of the replicative enzymes with the replication origins. It needs to be stated straight away that this is far from being understood. In the replication of simian virus 40 in mammalian cells, the virus-coded T-antigen binds to a specific group of sequences and then acts as a helicase to promote DNA strand separation [36]. The complex of host cell-encoded enzymes also contains a DNA-binding protein which specifically recognises an A-T-rich sequence within the SV40 replication origin and this binding event appears to be needed for efficient replication of the viral DNA by the complex [48]. A similar DNA-binding protein which recognises mammalian DNA replication origins has been isolated from HeLa cells [21]. The replication complex of yeast also contains a DNA binding activity and careful mapping in electron micrographs of the position of binding of the replication complex to yeast plasmids strongly suggests that the binding activity recognises the replication origins [33]. The more recent purification from yeast of individual proteins which bind replication origins [16, 17, 22] is further evidence that recognition of origins by a specific protein is involved in the initiation DNA replication. But what of plants? Unfortunately the evidence is very sparse and is confined to the observations made on the DNA polymerase multi-enzyme complex of pea. The multi-protein complex exhibits DNA binding activity [1, 15] which has been shown to reside in three different polypeptides of 118, 65 and 42 kDa, of which the 42 kDa is the most active. This polypeptide has been purified to homogeneity ([1]; J Al-Rashdi, SK Burton and JA Bryant, unpublished). The sequence specificity (if any) of the purified protein is now being investigated using a range of DNA sequences, including defined origins of replication, as substrates. The data obtained in these experiments will help to determine whether this DNA-binding protein functions as an origin recognition factor.

6. Regulation of enzyme activity within the cell cycle

The restriction of DNA replication to one particular phase of the cell cycle means that the enzymes of DNA replication are not needed for a significant proportion of each cycle. There is evidence from S. cerevisiae (see earlier) that the expression of genes coding for DNA replication enzymes fluctuates during the cell cycle although in S. pombe, only the one example is known (see earlier). There is also evidence from mammals for a cell cycle-dependent fluctuation in the expression of the primase gene, leading to fluctuations within the cell cycle of the population of polymerase-α which is associated with primase, but not in the total polymerase-α population [39]. Similarly in mammals, the expression of the PCNA gene peaks in S-phase [46], leading to changes in the amounts of PCNA-associated DNA polymerase δ, but not in the total polymerase δ population. Again, data from the plants are very scarce; the only really clear data concern PCNA, where expression of the gene fluctuates during the cell cycle, with a clear peak in the S-phase [38].

In addition to fluctuations in the amounts of enzyme regulated by changes in gene transcription, regulation could be based on association of the enzymes with the DNA template. Although the evidence for this is not extensive, it does appear in yeast [64] and in mammals [61] that polymerase-α is more tightly bound within the nucleus during S-phase than during other phases of the cell cycle. An extension of this observation is the suggestion that the formation of the replication complex from its constitutive enzymes occurs only in the S-phase and that this process is linked with the binding of the replicative enzymes within the nucleus [56]. Unfortunately it has proved very difficult to obtain clear evidence for this, and for now it must remain as just an interesting idea.

Finally we must consider the regulation of enzymes by post-translational phosphorylation. In animals, DNA polymerase-α, topoisomerase I and topoisomerase II are all phosphoproteins [28, 20, 40, 53, 55, 58]. For each of these enzymes, the phosphorylated form is very much more active than the dephosphorylated form. In plants, there is no information on the phosphorylation status of the DNA polymerases. However, it is known that a single-stranded binding protein from Lillium [32] and both topoisomerase I and topoisomerase II from maize are phosphoproteins [18]. For all three enzymes, dephosphorylation causes a significant loss of activity.

In addition to these specific instances of protein phosphorylation, there is also more general evidence that phosphorylation may be involved in the regulation of DNA replication. For example, the DNA polymerase complex of pea contains at least two protein kinase activities (P Brusa, M Crosti, JA Bryant and PN Fitchett, unpublished), one of which appears to reside in the same 42 kDa polypeptide as the DNA-binding activity (PN Fitchett, SK Burton and JA

Bryant, unpublished). Further, it is very clear that as meristem cells in the roots of germinating peas prepare to enter the S-phase, the activity of at least two protein kinases increases and several nuclear proteins become phosphorylated [see D Chiantante, this volume].

7. Relationship to plant development

The processes of cell division (including DNA replication) and DNA endoreduplication are developmentally regulated, leading to a defined spatial distribution within the plant of cells in which DNA replication or endoreduplication is occurring. Control at this level is of course only a feature of multicellular organisms, and for all the excellence of 'model' unicellular organisms, they cannot be specifically informative on this topic. There are also specific phases in plant development in which cells undergo a transition from quiescence to active proliferation (e.g. in germination) or from active proliferation to senescence (e.g. during the onset of dormancy). If the occurrence and location of DNA replication are controlled, what of the enzymes involved? Assays of enzyme activity and/or of enzyme protein indicate clearly that the presence of certain enzymes is strongly correlated with active cell proliferation and/or DNA endoreduplication. These include DNA polymerase-α, ribonuclease-H, DNA ligase, topoisomerase, PCNA and at least two protein kinases, including the p34[cdc2] protein kinase (see PCL John et al. and D Chiantante, this volume).

The development of in situ nucleic acid hybridisation and antibody probing techniques is now bringing a greater degree of resolution to these studies. It is now possible to ascertain, within a growing root or shoot, the distribution of particular mRNAs and proteins in relation to the distribution of DNA replication, cell division and endoreduplication. For the latter phenomenon it would be very interesting to study the distribution of p34[cdc2] and associated proteins in view of their role in linking S and M in S. pombe (see J Hayles and P Nurse, this volume). These investigations will facilitate the identification of the levels of control at which the activities of DNA replication proteins are regulated.

References

1. Al-Rashdi J, Burton SK and Bryant JA (1991) Characterization of DNA-binding activity from the DNA replication complex of pea (Pisum sativum). J Exp Bot 42: suppl, 49.
2. Araki H, Hamatake RK, Johnston LH and Sugino A (1991) DPB2, the gene encoding DNA polymerase II subunit B, is required for chromosome replication in Saccharomyces cerevisiae. Proc Natl Acad Sci USA 88: 4601-4605.
3. Araki H, Hamatake RK, Morrison A, Johnson AL, Johnston LH and Sugino A (1991) Cloning DPB3, the gene encoding the third subunit of DNA polymerase II of Saccharomyces cerevisiae. Nucleic Acids Res 19: 4867-4872.

4. Araki H, Ropp PA, Johnson AL, Johnston LH, Morrison A and Sugino A (1992) DNA polymerase II, the probable homolog of mammalian DNA polymerase epsilon, replicates chromosomal DNA in the yeast *Saccharomyces cerevisiae*. EMBO J 11: 733-740.

5. Bahman AM, Buck V, White A and Rosamond J (1988) Characterisation of the *CDC*7 product of *Saccharomyces cerevisiae* as a protein kinase needed for the initiation of DNA synthesis. Biochim Biophys Acta 951: 335-343.

6. Bambara RA and Jesse CB (1991) Properties of DNA polymerases δ and ε and their roles in eukaryotic DNA replication. Biochim Biophys Acta 1088: 11-24.

7. Bauer GA and Burgers MJ (1990) Molecular cloning, structure and expression of the yeast proliferating cell nuclear antigen gene. Nucleic Acids Res 18: 261-265.

8. Blow JJ (1988) Eukaryotic DNA replication reconstituted outside the cell. BioEssays 8: 149-152.

9. Blow JJ and Laskey RA (1988) A role for the nuclear envelope in controlling DNA replication within the cell cycle. Nature 332: 546-548.

10. Boulet A, Simon M, Faye G, Bauer GA and Burgers PMJ (1989) Structure and function of the *Saccharomyces cerevisiae CDC*2 gene encoding the large subunit of DNA polymerase III. EMBO J 8: 1849-1854.

11. Brill SJ and Stillman B (1991) Replication factor-A from *Saccharomyces cerevisiae* is encoded by three essential genes coordinately expressed at S phase. Genes Dev 5: 1589-1600.

12. Bryant JA (1992) Biochemical regulation of DNA replication. In: Biochemical Mechanisms Involved in Growth Regulation. C Smith and D Chiatante (eds), in press. London: Academic Press.

13. Bryant JA and Dunham VL (1988) Replication of nuclear DNA. Oxford Surveys Plant Molec Cell Biol 5: 23-55.

14. Bryant JA and Dunham VL (1988) DNA replication in plants. Boca Raton, Fl: CRC Press.

15. Bryant JA, Fitchett PN, Hughes SG and Sibson DR (1992) DNA polymerase-α in pea is part of a large multiprotein complex. J Exp Bot 43: 31-40.

16. Buchman AR, Kimmerly WJ, Rine J and Kornberg RD (1988) Two DNA-binding factors recognize specific sequences at silencers, upstream activating sequences, autonomously replicating sequences, and telomeres in *Saccharomyces cerevisiae*. Molec Cell Biol 8: 210-225.

17. Campbell J (1988) Eukaryotic DNA replication: yeast bares its ARS's. Trends Biochem Sci 13: 212-217.

18. Carballo M, Gine R, Santos M and Puigdomerech P (1991) Characterization of DNA topoisomerase II from cauliflower inflorescences. Plant Molec Biol 6: 137-144.

19. Chiatante D, Bryant JA and Fitchett PN (1991) DNA topoisomerase in nuclei purified from root meristems of *Pisum sativum*. J Exp Bot 42: 813-820.

20. Cripps-Wolfman J, Henshaw EC and Bambara RA (1989) Alterations in the phosphorylation and activity of DNA polymerase-α correlate with the change in replicative DNA synthesis as quiescent cells re-enter the cell cycle. J Biol Chem 264: 19478-19486.

21. Dailey L, Caddle MS, Heintz N and Heintz NH (1990) Purification of RIP60 and RIP100, mammalian proteins with origin-specific DNA-binding and ATP-dependent helicase activities. Molec Cell Biol 10: 6225-6235.

22. Diffley JFX and Stillman B (1990) The initiation of chromosomal DNA replication in eukaryotes. Trends Genet 6: 427-432.

23. Downey KM, Tan C-K, Andrews DM, Li X and So AG (1988) Proposed roles for DNA polymerases α and δ at the replication fork. In: Cancer Cells, vol 6, Eukaryotic DNA replication. T Kelly and B Stillman (eds), pp. 403-410. Cold Spring Harbor NY: Cold Spring Harbor Laboratory.

24. Fairman MP (1990) DNA polymerase δ/PCNA: actions and interactions. J Cell Sci 95: 1-4.

25. Fein K and Stillman B (1992) Identification of replication factor C from *Saccharomyces cerevisiae*: a component of the leading-strand DNA replication complex. Molec Cell Biol 12: 155-163.

26. Foiani M, Santocanale C, Plevani P and Lucchini G (1989) A single essential gene, *PRI*1,

encodes the large subunit of DNA primase in *Saccharomyces cerevisiae*. Molec Cell Biol 9: 3081-3087.

27. Francis D and Lyndon RF (1985) The control of the cell cycle in relation to floral induction. In: The Cell Division Cycle in Plants. J A Bryant and D Francis (eds), pp. 199-214. Cambridge, UK: Cambridge University Press.

28. Fry M and Loeb LA (1986) Animal Cell DNA Polymerases. Boca Raton, Fl: CRC Press.

29. Gordon CB and Campbell JL (1991) A cell cycle-responsive transcriptional control element and a negative control element in the gene encoding DNA polymerase alpha in *Saccharomyces cerevisiae*. Proc Natl Acad Sci USA 88: 6058-6062.

30. Hennessy KM, Clark CD and Botstein D (1991) Subcellular localization of yeast *CDC*46 varies with the cell cycle. Genes Dev 4: 2252-2263.

31. Hennessy KM, Lee A, Chen E and Botstein D (1991) A group of interacting yeast DNA replication genes. Genes Dev 5: 958-969.

32. Hotta Y and Stern H (1979) The effect of dephosphorylation on the properties of a helix-destabilizing protein from meiotic cells and its partial reversal by a protein kinase. Eur J Biochem 95: 31-38.

33. Jazwinski SM (1987) Evidence for involvement of a single major species of replicative complex in DNA synthesis from two diverse nuclear replicons in yeast. Biochem Biophys Res Commun 142: 1053-1058.

34. Johnson LM, Snyder M, Chang LMS, Davis RW and Campbell JL (1985) Isolation of the gene encoding yeast DNA polymerase I. Cell 43: 369-377.

35. Johnston LH and Lowndes NF (1992) Cell cycle control of DNA synthesis in budding yeast. Nucleic Acids Res 20: 2403-2410.

36. Kelly TJ (1988) SV40 DNA replication. J Biol Chem 263, 17889-17892.

37. Kitada K, Johnston LH, Sugino T and Sugino A (1992) Temperature sensitive *cdc*7 mutations of *Saccharomyces cerevisiae* are suppressed by the *DBF*4 gene, which is required for the G1/S cell-cycletransition. Genetics 131: 21-29.

38. Kodama H, Ito M, Ohnishi N, Suzaka I and Komamine A (1991) Molecular cloning of the gene for plant proliferating-cell nuclear antigen and expression of this gene during the cell cycle of *Catharanthus roseus* cells. Eur J Biochem 197: 459-503.

39. Kozu T, Seno T and Yagura T (1986) Activity levels of mouse DNA polymerase-α-primase complex (DNA replicase) and DNA polymerase-α, free from primase activity in synchronized cells and a comparison of their activities. Eur J Biochem 157: 215-259.

40. Kroll DJ and Rowe TC (1991) Phosphorylation of DNA topoisomerase II in a human cell line. J Biol Chem 266: 7957-7961.

41. Lowndes NF, Johnson AL, Breeden L and Johnston LH (1992) SWI6 protein is required for transcription of the periodically expressed DNA synthesis genes in budding yeast. Nature 357: 505-508.

42. Lowndes NF, Johnson AL and Johnston LH (1991) Coordination of expression of DNA synthesis genes in budding yeast by a cell-cycle regulated *trans* factor. Nature 350: 247-250.

43. Lowndes NF and Johnston LH (1992) Parallel pathways of cell cycle-regulated gene expression. Trends Genet 8: 79-81.

44. Lowndes NF, McInerny CJ, Johnson AL, Fantes PA and Johnston LH (1992) Control of DNA synthesis genes in fission yeast by the cell-cycle gene *cdc*10$^+$. Nature 355: 449-453.

45. Lucchini G, Francesconi S, Foiani M, Badaracco G and Plevani P (1987) The yeast DNA polymerase-primase complex: cloning of *PRI*1, a single essential gene related to DNA primase activity. EMBO J 6: 737-742.

46. Lui YC, Marrceino RL, Keng PC, Bambara RA, Lord EM, Chou WG and Zain SB (1989) Requirement for proliferating cell nuclear antigen expression during stages of the Chinese hamster ovary cell cycle. Biochemistry 28: 2967-2974.

47. Maine GT, Sinha P and Tye B-K (1984) Mutants of *S. cerevisiae* defective in the maintenance of minichromosomes. Genetics 106: 365-385.

48. Malkas LH and Baril EF (1989) Sequence recognition protein for the 17-base-pair A + T-rich

tract in the replication origin of simian virus 40 DNA. Proc Natl Acad Sci, USA 86: 70-74.

49. Malkas LH, Hickey RJ, Li C, Pedersen N and Baril EF (1990) A 21S enzyme complex from HeLa cells that functions in simian virus 40 DNA replication. Biochemistry 29: 6362-6374.

50. McIntosh EM, Atkinson T, Storms RK and Smith M (1991) Characterization of a short, *cis*-acting DNA sequence which conveys cell cycle stage-dependent transcription in *Saccharomyces cerevisiae*. Molec Cell Biol 11: 329-337.

51. Morrison A, Araki H, Clark AB, Hamatake RK and Sugino A (1990) A third essential DNA polymerase in *S. cerevisiae*. Cell 62: 1143-1151.

52. Nagl W, Pohl J and Radler A (1985) The DNA endoreduplication cycles. In: The Cell Division Cycle in Plants. JA Bryant and D Francis (es), pp. 217-232. Cambridge, UK: Cambridge University Press.

53. Nasheuer HP, Moore A, Wahl AF and Wang TSF (1991) Cell cycle-dependent phosphorylation of human DNA polymerase-α. J Biol Chem 266: 7892-7903.

54. Osley MA (1991) The regulation of histone synthesis in the cell cycle. Ann Rev Biochem 60: 827-861.

55. Pommier Y, Kerrigan D, Hartman KD and Glazer RI (1990) Phosphorylation of mammalian DNA topoisomerase I and activation by protein kinase C. J Biol Chem 265: 9418-9422.

56. Reddy GPV and Pardee AB (1980) Multi-enzyme complex for metabolic channeling in mammalian DNA replication. Proc Natl Acad Sci, USA 77: 3312-3316.

57. Richard MC, Litvak S and Castroviejo M (1991) DNA polymerase B from wheat embryos: a plant δ-like DNA polymerase. Arch Biochem Biophys 287: 141-150.

58. Saijo M, Enomoto T, Hanaoka F and Ui M (1990) Purification and characterization of type II DNA topoisomerase from mouse FM3A cells: phosphorylation of topoisomerase II and modification of its activity. Biochemistry 29: 583-590.

59. Salah-ud-Din, Brill SJ, Fairman MP and Stillman B (1990) Cell-cycle-regulated phosphorylation of DNA replication factor A from human and yeast cells. Genes Dev 4: 968-977.

60. Sitney KC, Budd ME and Campbell JL (1989) DNA polymerase III, a second essential DNA polymerase, is encoded by the *S. cerevisiae CDC*2 gene. Cell 56: 599-605.

61. Stokke T, Erikstein B, Holte H, Funderud S and Steen HB (1991) Cell cycle-specific expression and nuclear binding of DNA polymerase-α. Molec Cell Biol 11: 3384-3389.

62. Suzuka I, Daidoji H, Matsuoka M, Kodowaki K-I, Takasaki Y, Nakane PK and Moriuchi T (1989) Gene for proliferating cell nuclear antigen (DNA polymerase-δ auxiliary protein) is present in both mammalian and higher plant genomes. Proc Natl Acad Sci, USA 86: 3189-3193.

63. Suzuka I, Hata S, Matsuoaka M, Kosugi S and Hashimoto J (1991) Highly conserved structure of proliferating cell nuclear antigen (DNA polymerase-δ auxiliary protein) gene in plants. Eur J Biochem 197: 571-575.

64. Tsuchiya E, Kimura K, Miyakawa T and Fukui S (1984) Characteristic alteration in the nuclear polymerase activity during the cell division cycle of *Saccharomyces cerevisiae*. Nucleic Acids Res 2: 3143-3154.

65. White JHM, Barker DG, Nurse P and Johnston LH (1986) Periodic transcription as a means of regulating gene expression during the cell cycle: contrasting modes of expression of DNA ligase genes in budding and fission yeast. EMBO J 5: 1705-1709.

66. Yan H, Gibson S and Tye B-K (1991) *MCM*2 and *MCM*3, two proteins important for ARS activity, are related in structure and function. Genes Dev 5: 944-957.

5. Control of initiation of DNA replication in plants

MARK R.H. BUDDLES, MARCUS J. HAMER, JOHN ROSAMOND and
CLIFFORD M. BRAY

Abstract

Seed development, maturation and germination are developmental stages in the
life cycle of plants which are accompanied by programmed transitions from cell
proliferation to quiescence during the maturation phase and from quiescence to
reinitiation of cell proliferation in meristematic centres upon seed germination.
These transitions can be monitored readily in germinating wheat embryos and
are manipulated artificially in seed presowing treatments such as
osmoconditioning. In an attempt to investigate the molecular events controlling
the initiation of DNA synthesis at the G1/S-transition during the mitotic cell
division cycle in plants we have commenced a study into whether a homologue
of the budding yeast *CDC7* cell cycle control gene exists in plant tissues. The
yeast (*S. cerevisiae*) *CDC7* gene has been overexpressed in *E. coli*,
overproduced *CDC7* protein purified and subsequently used to prepare
polyclonal antibodies from rabbits. Protein extracts from germinating seeds or
meristematic tissues of several plant species including wheat, maize, leek and
Arabidopsis contain proteins of molecular mass $\sim$ 60 kDa which cross-react
with the *CDC7* polyclonal antibodies. Nuclei from these plant tissues are an
enriched source of the potential *CDC7* homologue. Making use of PCR and the
unique arrangements of the phosphoreceptor domain of the *CDC7* gene we
synthesise a PCR product having the expected size for use as a probe to screen
plant cDNA libraries for full length *CDC7* homologues. Initial experiments
involving functional complementation studies to isolate wheat *CDC7*
homologues have identified two clones of interest from a wheat cDNA λMax-1
yeast expression library. These clones are currently being characterised.

1. Introduction

An inherent ability to control the cell division cycle is of major importance
in the control of growth and differentiation in plants. This control is
exemplified during the transitions between periods of rapid growth and cell
division and cessation of these processes as seen during seed development, seed
germination, leaf development in cereals and floral induction in the shoot
apex. However, despite the obvious importance of these control mechanisms,
very little information is available at the molecular level concerning those

57

J.C. Ormrod and D. Francis (eds.), Molecular and Cell Biology of the Plant Cell Cycle, 57–74.
© 1993 *Kluwer Academic Publishers. Printed in the Netherlands.*

elements which regulate the cell division cycle in plants. An understanding at this level of those events which are crucial to the regulation of cell division would potentially allow a rational approach to be made towards the controlled modification of crop plant performance and also to the growth of plant cells in culture.

Cell division in higher plants is generally confined to meristematic regions and to a great extent plant morphogenesis is ultimately dependent on temporal and spatial control of cell division and cell expansion [8]. Although major control points within the mitotic cell cycle are thought to exist at the G1/S and G2/M-transitions [30] the identification of those biochemical components involved in plant cell cycle regulation has proved elusive. In contrast to the situation in plants, extensive biochemical and genetic approaches have been used successfully in the study of the cell division cycle in the budding yeast *Saccharomyces cerevisiae* and the fission yeast *Schizosaccharomyces pombe*. In these lower eukaryotes major control points have been demonstrated to exist at a point termed START located in G1 at the point where the cell becomes committed to the mitotic cell cycle and which is analogous to the 'R' control point in vertebrate cells, and at a second point located in G2 which determines the initiation of mitosis [24].

More recently it has been demonstrated that at least some of the cell division cycle regulatory molecules are highly conserved amongst eukaryotic organisms and this has permitted the functional complementation of mutations in yeast cell division cycle (*cdc*) genes by homologues from distantly related organisms including man [20] and higher plants [11, 14]. Additionally, at least one of the proteins responsible for the coupling of DNA replication to chromosome segregation at mitosis is the *cdc2/CDC28* (p34) protein kinase which appears conserved in a range of unrelated organisms [7]. The p34^{cdc2} protein kinase also seems to have an important role to play at START in addition to its role at the G2/M-transition [23]. In contrast to the extensive studies on control mechanisms operative at the G2/M transition, much less is known about the biochemical controls operating at the other major control point in the mitotic cell cycle when the cell becomes committed to progress through the cell cycle late in G1-phase at START. Plant cells may arrest growth in either G1 or G2 depending on species, tissue or developmental growth stage. Our interest in the control mechanisms operative in the mitotic cell division cycle at START arose from our long term interest in seed germination, viability and vigour combined with several observations detailing the arrest of meristematic cells during seed development in the G1-phase of the cell cycle [2]. This arrest in G1 necessitates a commitment of these cells to a round of DNA replication and hence passage through START upon the reinitiation of cell division during seed germination. The efficient reinitiation of cell division especially in the embryo or embryonic axis of the seed during germination appears to relate to the vigour rating or quality of the seed lot [5].

In *Saccharomyces cerevisiae* the *CDC7* gene appears to have multiple roles in cell physiology being required for the initiation of mitotic DNA synthesis,

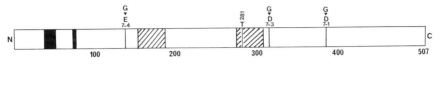

Fig. 1. The S. cerevisiae CDC7 protein kinase.

necessary for genetic recombination during meiosis and the formation of ascospores and has also been implicated in an error-prone DNA repair pathway [26]. The yeast chromosomal *CDC7* gene has been isolated and characterised and its protein product, p58^{CDC7}, shown to encode a serine/threonine protein kinase [1, 15]. This protein kinase activity is essential for its function in both mitosis and meiosis and the kinase activity *per se* is potentially regulated by phosphorylation of the p58^{CDC7} protein [6]. The *CDC7* protein kinase is unique amongst all serine/threonine protein kinases sequenced to date in that the highly conserved residues of its functional catalytic domain differ markedly in from that seen in all other known serine/threonine protein kinases (Fig. 1). This unique organisation is essential for the cellular function of the *CDC7* gene product [1]. The p58^{CDC7} protein kinase functions immediately prior to the initiation of DNA synthesis at the G1/S transition in the mitotic cell cycle acting downstream of the *cdc2/CDC28* kinase function at START [15] and occupies a key position in the transition from G1 into the S-phase of the mitotic cell division cycle.

Since signficant success had been achieved via a combination of approaches in the isolation of plant homologues of the *cdc2/CDC28* yeast cell cycle control genes [10, 11] it was decided to use a combination of immunological and molecular genetic approaches to investigate whether a *CDC7* homologue existed in plant tissues and to isolate and characterise this homologue. If this approach was successful then the longer term aim would be to determine the function of the plant *CDC7* homologue in the control of the mitotic cell division cycle, meiosis and DNA repair in plant cells. This report describes the initial steps towards characterisation of a such a wheat *CDC7* homologue.

2. Materials and methods

Plant materials and germination conditions

Wheat (*Triticum aestivum* var *Galahad*) seed was supplied by Quantil Seeds Ltd, Ormskirk, Lancs, U.K. Wheat embryos were prepared, germinated and RNA extracted as previously described [4]. Primary wheat leaves were prepared from wheat seedlings grown in a controlled environment cabinet for 7d [31]. Leek seeds (*Allium porrum* L. cv Verina) were obtained from Breeders Seeds Ltd, Ormskirk. Lancs, UK, and were primed and germinated as described previously [4]. Maize and *Arabidopsis* seeds were obtained through the AFRC PMB programme and plants were grown in a controlled environment cabinet under conditions identical to those for 7dd wheat seedlings.

2.2. Estimation of rates of DNA synthesis

Incorporation of methyl-[^{3}H] thymidine into DNA in pulse-labelling experiments with wheat and leek embryos was performed as described previously [3, 5].

2.3. DNA procedures, bacterial and yeast manipulations

These procedures were performed using established protocols [22] unless otherwise stated.

2.4. Antibody production

The coding sequence of the *S. cerevisiae CDC7* gene was subcloned as an NcoI-SalI fragment into the plasmid pJLA502 (Fig. 2) to produce the plasmid pJLA272 where expression of *CDC7* was under the control of the thermoinducible $\lambda P_L P_R$ promoter. The protease deficient *E. coli* strain CAG597 was transfected with the plasmid pJLA502 (control) or pJLA272 and *CDC7* overexpression induced in a mid log phase culture grown at 30°C by a 2 h heat shock at 42°C. After heat shock, cells were collected by centrifugation and the bacterial pellet resuspended in Laemmli sample buffer prior to electrophoresis on 10% SDS-polyacrylamide gels using a Biorad minigel system [19].

Inclusion bodies were prepared from heat shocked bacterial cells by freezing the bacterial pellet at −70°C for 30 min, resuspending the thawed pellet in 50 mM Tris-HCl buffer pH 8.0 containing 25% (w/v) sucrose, 1 mM EDTA followed by the addition of lysozyme to lyse the cells. DNA was digested by addition of DNase I and incubation for 30 min on ice. Inclusion bodies were precipitated by centrifugation at 5000 g for 10 min at 4°C, washed with 0.5% Triton X-100, 1 mM EDTA (twice) and finally solubilised in 8M urea. Solubilised inclusion body proteins were separated on 10% SDS-poly-

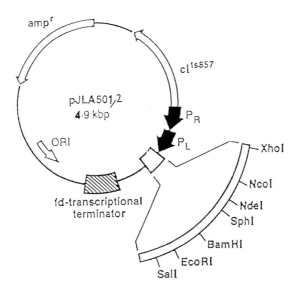

Fig. 2. Schematic diagram of the plasma pJLA 502 showing the multiple cloning site. The coding sequence of the *S. cerevisiae CDC7* gene was cloned into the NcoI/SalI sites.

acrylamide gels as described previously.

To obtain sufficient p58^{CDC7} for antibody production, solubilised inclusion body proteins were separated by preparative SDS-PAGE (10% gels), p58^{CDC7} fractions electroeluted from the gel and dialysed against 3 changes of double distilled water before freeze-drying. The lyophilised material was redissolved in 1 cm^3 sterile distilled water and the protein concentration determined [21]. Between 1–2 mg of p58^{CDC7} protein could be purified in this way.

Purified p58^{CDC7} (500μg) in complete Freunds adjuvant was used to inject rabbits subcutaneously followed by a second injection after 2–3 weeks. A sample of preimmune serum was taken from rabbits prior to the first injection followed by test bleeds from an ear vein at regular intervals. Serum was tested against a cell extract from heat shocked *E. coli* (CAG597 containing pJLA272 or pJLA502) via SDS-PAGE and Western blotting. When a good response was obtained (2–3 weeks after the second injection) blood samples (70 cm^3) were obtained from the rabbits, the blood allowed to clot overnight at 4°C, centrifuged at 10,000 g for 10 min at 4°C and the serum removed from the clotted blood. The serum was stored at −70°C until use and used at a dilution of 1 in 500.

2.5. Plant tissue sample preparation

Whole plant homogenates were prepared for gel electrophoresis by grinding tissue to a powder in liquid N$_2$ with a mortar and pestle followed by homogenisation in nuclear extraction buffer [17] in a glass teflon homogeniser and filtration of the homogenate through three layers of Miracloth. The clarified homogenate was assayed for protein [21] and 20 μg protein samples separated by SDS PAGE (10% gels) prior to staining gels for protein or transfer to nitrocellulose membranes. Western blotting using p58[CDC7] antisera was performed using the ECL Western blotting system in accordance with the manufacturers instructions (Amersham International plc, U.K.). Nuclear fractions were prepared from plant material using an established protocol [17].

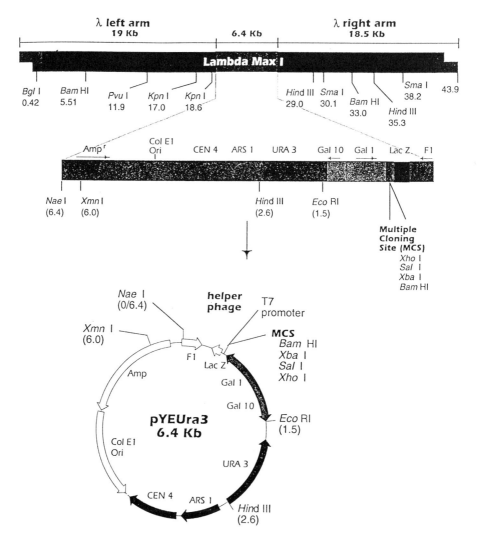

Fig. 3. Schematic diagram of Lambda Max1 and pYEUra3. The pYEUra3 sequences can be excised from the lambda Max1 and converted into plasmid by helper phage.

2.6. Polymerase chain reaction (PCR)

The touchdown PCR method was used in all PCR amplification studies [9] using degenerate oligonucleotides based upon two conserved regions IIHRDIKP and RAPEVL [12, 26] flanking the phosphoreceptor domain of yeast *CDC7* as forward and reverse primers.

2.7. Functional complementation studies

The cDNA expression library used in these studies was supplied by Clontech Laboratories, (Ca, USA) and prepared using mRNA from 24 h germinated wheat embryos to synthesise cDNA which was subsequently cloned into the EcoR1 site of the yeast expression vector λMax-1 so that expression of the cloned cDNA was under the control of the inducible yeast GAL10 promoter (Fig. 3). The plasmid pYEUra3 containing the cDNA library was excised from λMax-1 using helper phage in accordance with the manufacturers instructions and used to transform a temperature sensitive mutant of *S. cerevisiae*, strain SB 646 (*cdc7–3, ura 3–52, ade 2*) using a LiAc transformation protocol essentially as described [27]. Transformants were selected for their ability to grow in the presence of galactose at the restrictive temperature (35°C). Since yeast cells may harbour more than one plasmid, plasmid minipreps from yeast colonies growing at the restrictive temperature were used to transform *E. coli* (HW87) by electroporation. Minipreps of plasmids from single *E. coli* colonies which should contain homogeneous plasmid species were then used in LiAc transformation of *S. cerevisiae* SB646, when high frequency rescue for growth of yeast transformants at the restrictive temperature would be expected for plasmids containing *CDC7* functional homologues. Such plasmids containing inserts of interest were subjected to further analysis and characterisation.

3. Results

3.1. Cell division and replication of nuclear DNA in seeds

In mature 'dry' orthodox seeds the embryo or embryonic axis contains cells arrested both in G1 and G2 in some species whilst in other species arrest is exclusively in G1 [2]. Within a particular species the proportions of cells arrested in G1 and G2 varies between different parts of the seed [2]. Upon seed germination, replication of nuclear DNA and reinitiation of cell division are usually post-germinative events since the initial phase of radicle elongation associated with completion of germination occurs almost exclusively via cell elongation. This delay between completion of imbibition and the onset of replicative DNA synthesis lasts for about 10–16 h in wheat embryos (Fig. 4).

Osmopriming is a commercial pre-sowing seed treatment which enhances subsequent germination performance [5] and we have demonstrated that during

64

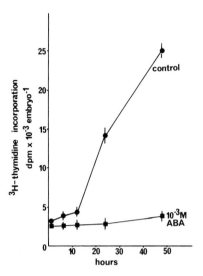

Fig. 4. DNA synthesis in wheat embryos germinated at 20°C in the presence or absence of 10^{-3}M abscisic acid, an inhibitor of replicative DNA synthesis, in the germination medium. Results represent the average + SEM of 3 independent experiments.

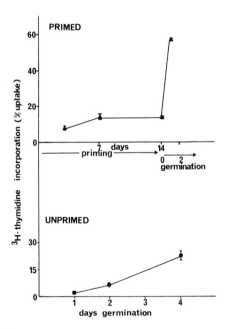

Fig. 5. DNA synthesis in leek embryo tissue during priming and/or germination at 15°C. Results represent the average of duplicate determinations. The level of incorporation is expressed as a percentage of the amount of radioactive precursor taken up by the embryo sample at each priming or germination point.

osmopriming of leek seeds there is no initiation of nuclear DNA replication or cell division in hydrated leek embryo tissue even over a priming period lasting for 14 days (Fig. 5 and D Gray, personal communication). However subsequent germination of osmoprimed leek seeds is accompanied by a rapid initiation of nuclear DNA replication and cell division (Fig. 4) but these processes appear dependent upon continued *de novo* protein biosynthesis during the immediate post-priming period (RM Taylor and CM Bray, unpublished results). At least one of the proteins which is synthesised during this postpriming period appears to be an aphidicolin-sensitive DNA polymerase (M Ashraf and CM Bray, unpublished results). We have subsequently concentrated our efforts on the identification of those factors controlling the reinitiation of DNA replication during the onset of growth processes during seed germination and seedling establishment and in particular we have commenced investigations to determine whether a homologue of the yeast *CDC7* gene exists in plant tissues.

3.2. Overexpression of the Saccharomyces cerevisiae CDC7 gene in E. coli

Protein samples from heat shocked *E. coli* (CAG597) cultures previously transformed with pJLA502 (control) or pJLA272 were analysed by SDS-PAGE (Fig. 6). An extra protein band could be detected on stained gels in the sample from the heat-shocked *E. coli* containing plasmid pJLA272; this protein species migrated with an apparent molecular mass of $\sim$ 54 kDa in SDS gels. No equivalent stained band was present in the protein sample from *E. coli* containing the plasmid pJLA502.

Overexpression of heterologous proteins in *E. coli* often results in the formation of inclusion bodies within the bacterial cell. These inclusion bodies are insoluble complexes which contain the overexpressed protein and bacterial cellular components. In the case of *E. coli* transformed with pJLA272, the overexpressed yeast *CDC7* protein was to be found complexed within inclusion bodies and was only effectively solubilised in the presence of 8M urea (Fig. 6). The urea-solubilised material from these inclusion bodies was an enriched source of p58^{CDC7} but when the solublised inclusion body protein was loaded onto SDS polyacrylamide gels, the p58^{CDC7} protein migrated anomalously through the gel with an apparent molecular mass of 47 kDa (Fig. 6). Purified p58^{CDC7} protein prepared by elution from preparative polyacrylamide gels was used in the preparation of polyclonal antibodies from rabbits.

3.3. Use of polyclonal antibodies to detect p58^{CDC7} homologues in plant tissues

Using the polyclonal antibodies prepared against *S. cerevisiae* p58^{CDC7} combined with Western blotting techniques protein extracts prepared from germinating embryos or seedlings of wheat (*Triticum aestivum* L.), maize (*Zea mays*), leek (*Allium porrum* L.) and *Arabidopsis thaliana* contained a protein

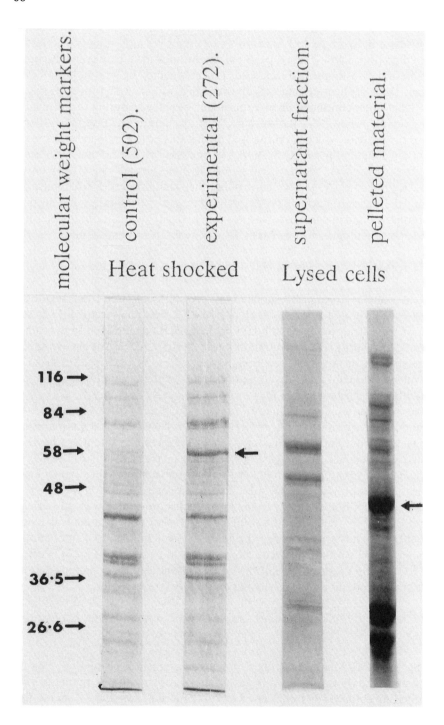

Fig. 6. Overexpression of p58^{CDC7} after cloning into the plasmid pJLA 502.

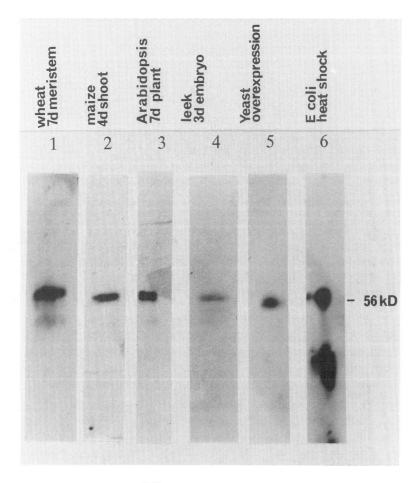

Fig. 7. Western blot showing p58^CDC7 homologues in plant tissues.

species migrating with an apparent molecular mass of approximately 60 kDa which cross-reacted with the *cdc*7 polyclonal antibodies (Fig. 7). Nuclear fractions from these plant tissues were an enriched source of this potential p58^CDC7 homologue.

3.4. CDC7 homologues in meristematic and non-meristematic tissues of the developing wheat leaf

Young leaves of grasses exhibit an ordered array of cells at various stages of differentiation ranging from meristematic cells at the base of the leaf through a smooth gradient of differentiating non-dividing cells to highly differentiated cells at the leaf tip [30]. The well-defined developmental status of cells at specific positions within the young primary leaf allows this tissue to be used as a convenient model system to investigate the possible relationship between the

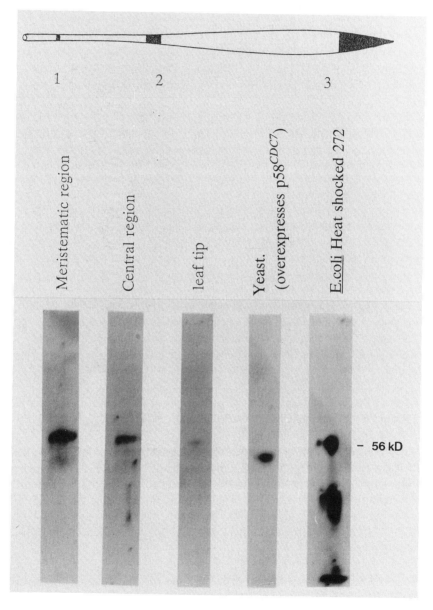

Fig. 8. Localisation of p58CDC7 immunopositive material in the primary wheat leaf.

potential of the leaf cells to proliferate and the level of p58^{CDC7} protein within dividing and non-dividing terminally differentiated cells.

Immunopositive material reacting with the polyclonal antibodies to p58^{CDC7} was detected in all three areas of the primary wheat leaf taken for analysis i.e. meristematic tissue at the base of the leaf, differentiated non-dividing tissue midway between the leaf base and the leaf tip, and terminally differentiated non-dividing tissue at the leaf tip (Fig. 8). However, the level of the protein species of molecular mass 60 kDa cross reacting with the polyclonal antibodies to p58^{CDC7} appears to decline as cells progress from the mersitematic regions at the leaf base through to non-mersitematic cells of the leaf mid region and leaf tip (Fig. 8). A densitometric scan of the Western blot showed a relative intensity of p58^{CDC7} immunopositive material of 100 : 38 : 5 in meristematic, mid and leaf tip regions of the leaf, respectively. A similar pattern is seen whether leaf extracts are loaded onto the gel on the basis of equivalent amounts of protein present in the leaf extracts from the different regions of the leaf or as protein loadings calculated to equate to equivalent DNA contents for each wheat leaf extract preparation.

3.5. PCR amplification

Comparison of sequence data on all known protein kinases has revealed the presence of several highly conserved regions [12]. Two such amino acid sequences flanking the unique arrangement of the phosphoreceptor domain of the *S. cerevisiae CDC7* protein kinase (Fig. 9) were used as a basis to synthesise primers for PCR with DNA prepared from wheat λMax-1, wheat λgt11 and *Arabidopsis* λgt11 cDNA libraries. After PCR, a band of the expected size

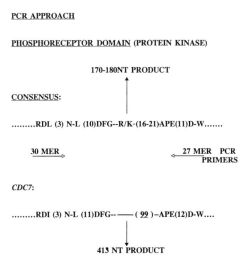

Fig. 9. The *Saccharomyces cerevisiae CDC7* protein kinase phosphoreceptor domain indicating the positions of the primers used in PCR amplification reactions.

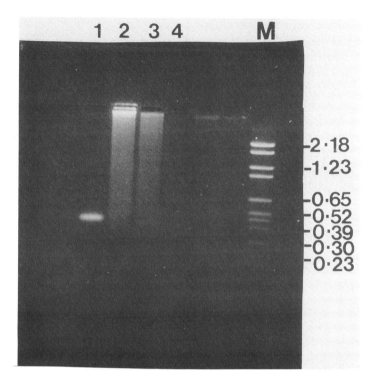

Fig. 10. PCR product analysis. Analysis of 'touchdown' PCR products showing: lane 1 pVB401, a *CDC*7 containing plasmid used as a positive control; lane 2 PCR products using a wheat λMax cDNA library; lane 3 PCR products using a wheat λgt11 cDNA library; lane 4 PCR products using an *Arabidopsis* λgt11 cDNA library.

(~ 420 bp) was amplified from the wheat λgt11 library (Fig. 10). Experiments are now in progress to vary the PCR conditions to optimise production of this ~ 420 bp band prior to sequence characterisation and subsequent use as a probe in isolation of a full length cDNA clone of the putative plant *CDC*7 homologue from existing cDNA libraries.

3.6. *Functional complementation of a cdc7 mutation of S. cerevisiae*

The ability to complement a *cdc*7 mutation of *S. cerevisiae* would be the definitive proof of the functionality of a *CDC*7 homologue. We have used the wheat cDNA library in the yeast expression vector λMax-1 to transform a temperature-sensitive *S. cerevisiae cdc* 7–3 strain and screened for the ability of transformants to grow at 35°C. Several transformants able to grow at 35°C have been isolated and of these two clones of interest have been demonstrated to contain pYEUra3 plasmids having 2 kb and 750 bp inserts respectively (Fig. 11). These clones are currently being characterised by DNA sequencing and by their ability to permit allele-specific rescue of other *cdc*7 *ts* mutations.

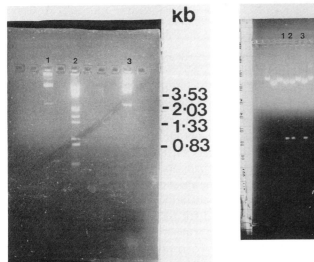
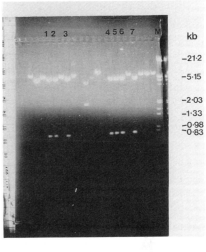

Fig. 11. Agarose gel electrophoresis of *Eco*RI digestion products of pYEUra3 recombinant plasmids which permit high frequency rescue of the t_s *cdc 7–3* mutant allele of *S. cerevisiae* at 35°C (a) showing 2 kb inserts in lanes 1 and 3; λDNA *Hin*d III/*Eco*RI digestion products used as markers are shown in lane 2 (b) showing 750 bp inserts in lanes 1–7 and markers consisting of a mixture of fragments from *Bg*lI digested pBR328 and *Hinf*I digested pBR328 in lane M.

4. Discussion

During the G1-phase of the cell cycle, cells prepare for S- phase which is marked by the commencement of DNA, histone and some enzyme synthesis ([25], see SJ Aves and JA Bryant, this volume). Extracellular factors also determine whether a quiescent cell will begin to proliferate and also whether a normal proliferating cell in the G1-phase of the cell cycle will become quiescent or progress through the cell cycle to mitosis [25]. These controls are exemplified in plants by the changes in cell cycle progression seen during the different stages of seed development, maturation and germination which initiated our studies into factors controlling the progression from G1-into S-phase of the cell cycle and the control of initiation of DNA synthesis. Several recent reports have demonstrated that functional homologues of the *S. pombe/S. cerevisiae* *cdc2/CDC*28 gene can be isolated from plant tissues [7, 14] but have also indicated that the situation in plants may be more complex than that in yeasts since at least two cell cycle controlling *cdc2* gene homologues appear to be present in *Arabidopsis* [13]. Nevertheless at least some aspects of the higher plant cell cycle are probably controlled by a p34[cdc2] centreed regulatory system similar to that of yeasts with the *cdc2/CDC*28 gene products interacting with a different set of regulatory molecules at the two major control points in the mitotic cell division cycle i.e. at START and entry into mitosis [23]. Genes expressed preferentially at the G1/S boundary and in S-phase have been identified in synchronised cultures of *Catharanthus roseus* cells [16, 18] and the identification of their protein products is awaited with interest (also see T Nagata and Y Takahashi, this volume).

In the present study we have demonstrated that polyclonal antibodies raised against the budding yeast protein p58^{CDC7} recognise a protein species of ~ 60 kDa in several unrelated plant species. This potential plant homologue of p58^{CDC7} appears to be localised in the nucleus and to be present at higher levels in actively dividing cells than in terminally differentiated non-dividing cells. The continued presence of protein recognised by the p58^{CDC7} polyclonal antibodies in non-dividing plant cells in the absence of replicative nuclear DNA synthesis may be indicative of a post-translational modification of p58^{CDC7} homologue activity as has been suggested for p58^{CDC7} in yeast [6] during the mitotic cell cycle. This contrasts with the transcriptional regulation of *CDC7* gene expression during meiosis where *CDC7* transcript levels vary, reaching a maximum near the time at which recombination occurs [28].

One reservation concerning the use of antibodies as the only route to the identification of a plant *CDC7* homologue is the possibility that the polyclonal antibodies may not be recognising 'authentic' plant *CDC7* protein but instead may recognise a closely related protein which has no functional equivalence. Nevertheless, this antibody approach has recently proved successful in the isolation of a pea *cdc2* homologue [10]. However, we have also begun to explore alternatives to the antibody approach by using PCR and functional complementation of appropriate yeast temperature sensitive *cdc7* mutants with which to isolate a plant *CDC7* homologue. Both these approaches have been used successfully in the isolation of plant *cdc2* homologues from *Arabidopsis* [11], maize [7] and alfalfa [14]. As discussed previously, the unique arrangement of the conserved phosphoreceptor domain of *CDC7* (Fig. 9) makes the PCR approach an attractive one since this unique sequence arrangement is essential for p58^{CDC7} activity in initiation of DNA replication and the use of oligonucleotide primers to conserved sequences flanking this region would be expected to produce unique PCR products. Preliminary experiments have shown this to be the case (Fig. 10) although optimisation of PCR protocols has still to be performed prior to use of this ~ 420 bp product in library screening studies. The functional complementation approach has required us to construct appropriate plant cDNA libraries in yeast expression vectors and also suitable yeast strains for use in these complementation studies. These are now available along with optimised protocols which yield transformation frequencies 10-fold greater than with existing PEG and LiCl transformation methods. Preliminary studies have already produced clones of potential interest (Fig. 11) which are currently being characterised.

In the budding yeast the activity of *CDC7* at the G1/S-phase boundary is dependent upon a number of other gene products which act in cells committed to mitosis (i.e post START) but prior to initiation of DNA synthesis. These genes include *CDC*4, *DBF*4, *ICD*1, *SCM*4, and *HFS*1 whose interactions may play a role in modulating nuclear/chromatin structure prior to S-phase. The successful completion of our initial objective of cloning the plant homologue of *CDC*7 will then provide the potential to extend significantly our understanding of the events which occur in the plant cell prior to the initiation of DNA synthesis by using the PCR or functional complementation approach to search for homologues of those genes whose products either interact with or modulate *CDC*7 activity during the cell cycle. These longer term aims are designed to help elucidate the molecular mechanisms by which the events at the G1/S-phase boundary are co-ordinated and so ultimately provide information on those controls which serve to activate the cell cycle during cell proliferation or return dividing cells to the quiescent state in various plant tissues at specific developmental stages.

Acknowledgements

We thank the AFRC (Plant Molecular Biology Programme) for financial support.

References

1. Bahman M, Buck V, White A and Rosamond J (1988) Characterisation of the CDC7 gene product of *S. cerevisiae* as a protein kinase needed for the initiation of mitotic DNA synthesis. Biochim Biophys Acta 951: 335–343.
2. Bewley JD and Black M (1985) Seeds: Physiology of development and germination. Plenum Press, New York.
3. Blowers LE, Stormonth DA and Bray CM (1980) Nucleic acid and protein synthesis and loss of vigour in germinating wheat embryos. Planta 150: 19–25.
4. Bray CM and Smith CAD (1985) Stored polyadenylated RNA and loss of vigour in germinating wheat embryos. Plant Science 38: 71–79.
5. Bray CM, Davison PA, Ashraf M and Taylor RM (1989) Biochemical changes during osmopriming of leek seeds. Annals Bot 63: 185–193.
6. Buck V, White A and Rosamond J (1991) CDC7 protein kinase activity is required for mitosis and meiosis in *S. cerevisiae*. Mol Gen Genet 227: 452–457.
7. Colasanti J, Tyers M and Sundaresan V (1991) Isolation and characterisation of cDNA clones encoding a functional p34^{cdc2} homologue from *Zea mays*. Proc Natl Acad Sci USA 88: 3377–3381.
8. Doonan J (1991) Cycling plant cells. The Plant Journal 1: 129–132.
9. Don RH, Cox PT, Wainwright BJ, Baker K and Mattick JS (1991) 'Touchdown' PCR to circumvent spurious priming during gene amplification. Nucleic Acids Res 19: 4008.
10. Feiler HS and Jacobs TW (1991) Cloning of the pea *cdc*2 homologue by efficient immunological screening of PCR products. Plant Mol Biol 17: 321–333.
11. Ferreira PCG, Hemerly AS, Villarroel R, Van Montagu M and Inze D (1991) The *Arabidopsis* functional homolog of the p34^{3cdc2} protein kinase. The Plant Cell 3: 531–540.
12. Hanks SK, Quinn AM and Hunter T (1988) The protein kinase family: Conserved features and deduced phylogeny of the catalytic domains. Science 241: 42–52.
13. Hirayama T, Imajuku Y, Anai T, Matsui M and Oka A (1991) Identification of two cell cycle controlling *cdc*2 gene homologs in *Arabidopsis thaliana*. Gene 105: 159–165.

14. Hirt H, Pay A, Gyorgyey J, Bako L, Nemeth K, Borge L, Schweyen RJ, Heberle-Bors E and Dudits D (1991) Complementation of a yeast cell cycle mutant by an alfalfa cDNA encoding a protein kinase homologous to p34^{cdc2}. Proc Natl Acad Sci USA 88: 1636–1640.
15. Hollingsworth RE and Sclafani RA (1990) DNA metabolism gene CDC7 from yeast encodes a serine (threonine) protein kinase. Proc Natl Acad Sci USA 87: 6272–6276.
16. Ito M, Kodama H and Komamine A (1991) Identification of a novel S-phase-specific gene during the cell cycle in synchronous cultures of *Catharanthus roseus* cells. The Plant Journal 1: 141–148.
17. Kalinsky A, Chandra GR and Muthukrishnan S (1986) Study of barley endonucleases and α-amylase genes. J Biol Chem 261: 11393–11397.
18. Kodama H, Ito M, Hattori T, Nakamura K and Komamine A (1991) Isolation of genes that are preferentially expressed at the G1/S boundary during the cell cycle in synchronised cultures of *Catharanthus roseus* cells. Plant Physiol 95: 406–411.
19. Laemmli UK (1970) Cleavage of structural proteins during the assembly of the head of the bacteriophage T$_4$. Nature 227: 680–685.
20. Lee MG and Nurse P (1987) Complementation used to clone a human homologue of the fission yeast cell cycle control gene *cdc2*. Nature 327: 31–35.
21. Lowry OH, Rosebrough NJ, Farr AL and Randall RJ (1951) Protein measurement with the folin phenol reagent. J Biol Chem 193: 265–300.
22. Maniatis T, Fritsch EF and Sambrook J (1982) Molecular cloning: A laboratory manual. Cold Spring Harbor, NY: Cold Spring Harbor Laboratory.
23. North G (1991) Starting and Stopping. Nature 351: 604–605.
24. Nurse P (1990) Universal control mechanism regulating onset of M-phase. Nature 344: 503–507.
25. Pardee AB (1989) G1 events and regulation of cell proliferation. Science 246: 603–608.
26. Patterson M, Sclafani RA, Fangman WL and Rosamond J (1986) Molecular characterisation of cell cycle gene CDC7 from *S. cerevisiae*. Mol Cell Biol 6: 1590–1598.
27. Schiestl RH and Gietz RD (1989) High efficiency transformation of intact yeast cells using single stranded nuclei acids as a carrier. Curr Genet 16: 339–346.
28. Sclafani RA, Patterson M, Rosamond J and Fangman WL (1988) Differential regulation of the yeast CDC7 gene during mitosis and meiosis. Mol Cell Biol 8: 293–300.
29. Van 't Hof J (1985) Control points within the cell cycle. In: JA Bryant and D Francis (eds) The cell division cycle in plants. pp. 1–15. Cambridge, Cambridge University Press, U.K.
30. Wernicke W and Milkovits L (1987) Effect of auxin on the mitotic cell cycle in cultured leaf segments at different stages of development in wheat. Physiol Plantarum 69: 16–22.

6. Nuclear proteins and the release from quiescence of root meristematic cells in *pisum sativum*

DONATO CHIATANTE

Abstract

Chromatin comprises DNA highly compacted with many proteins, and both transcription and replication processes depend on the molecular mechanisms which are able to modify this structural organization. The state of chromatin compaction is very high in cells of dry embryos and decreases with the imbibition which takes place during the first phase of germination. The structural nuclear proteins and the enzymes involved in DNA metabolism play an important role in these variations, therefore they might control the release from the quiescence of cells during germination. QP47 is a novel nonhistone nuclear protein which seems to have a structural role similar to that of H1. The variation of content of this protein could influence the accessibility of DNA.

1. Introduction

Usually, a cell is termed quiescent when it is unable to divide but is still viable. The lack of cell division does not necessarily involve loss of the accessibility of DNA; a non proliferating cell can maintain a part of its genome in a condition to be accessible for transcription while DNA replication is inhibited. This is the normal condition of differentiated cells even though they are never termed quiescent.

All the tissues forming the seed are called quiescent when the seed is still dry and during the initial stage of its imbibition. In this case, not only is DNA replication arrested but the chromatin is in a state of inactivity or "silence" with regard to any reaction involving DNA metabolism. This condition of the chromatin might suggest a block of the accessibility of DNA.

To avoid confusion when referring to the cell cycle condition in higher plants, we should perhaps use the term "arrested" for cells whose DNA replication is blocked but whose chromatin is active, and use the term "quiescent" only for the cells in dried embryos of the seed whose chromatin is completely inactive.

In these two cases, the use of a specific term would indicate the existence of differences of organization of the chromatin responsible for the state of accessibility to replication and transcription complexes. Such differences have not been investigated so far, and in this paper I shall consider some aspects of the organization of the chromatin in "quiescent" cells of the seed embryo.

J.C. Ormrod and D. Francis (eds.), Molecular and Cell Biology of the Plant Cell Cycle, 75–83.
© 1993 *Kluwer Academic Publishers. Printed in the Netherlands.*

A comparison with the chromatin organization of proliferating cells will also be made.

Particular attention will be paid to two factors which might control the accessibility of the chromatin and enforce (or release) the state of quiescence of the cells in the seed of *Pisum sativum*: the variations of structural organization of the chromatin due to the state of hydration of the tissues and the presence or absence of particular nuclear proteins (with enzymic or structural function).

A better knowledge of the mechanism of enforcement and release from quiescence of embryo cells, should throw light on the more complex process of germination.

2. Variations of structural organization of the chromatin in nuclei of quiescent or proliferating root meristems

A complex series of secondary metabolic events take place in the embryo cells in response to primary signals received from the seed (hormones, environmental conditions, etc, etc). The sum of such metabolic events induces the enforcement of, or the release from, quiescence in the embryo cells during seed maturation or germination. The variations of the structural organization of the chromatin correlated with the state of hydration of the tissues are certainly among those events and result in the modification of DNA compactness and accessibility, with obvious effects on all the other cellular activities.

In the literature, there are no reports regarding chromatin changes during the maturation of the seed. Hence there is a lack of information concerning the modification which the chromatin undergoes in nuclei of cells preparing to enter a state of quiescence. Probably, this is a consequence of the problems connected with the preparation of the fresh material from maturing seeds. On the other hand, very little information is available regarding the release from quiescence during germination. The dried nuclei of the embryo are characterized by a very compact chromatin with granular aggregations [1]; the nucleolus presents a dense fibrillar component [2]. When the DNA is removed by digestion, the nuclear matrix appears to be in a state of disorganization [11]. Moreover, it has been repeated that the size of dried nuclei is considerably smaller than that of imbibed nuclei [1].

From a functional point of view, these variations of structural organization of the chromatin during seed dehydration, cause severe damage to the DNA molecules and therefore a considerable amount of repairing activity must be carried out before the onset of DNA metabolism [18].

In proliferating cells, the hydration has been completed and nuclei have returned to their normal dimensions [1]. The chromatin has returned to its normal condition of dispersion and the nucleolus becomes vacuolated. An analysis of the nuclear matrix after hydrolysis of the DNA, shows that a normal matrix has been formed, with the appearance of a regular network of a fibrillar nature (the biochemical composition is unknown).

In the case of *Pisum sativum* it can be observed that the nuclei of the dry embryo comprise chromatin which is not well organized into euchromatic and heterochromatic zones and the nucleolus contains many vacuoles. After 72 h of imbibition, the nuclei consist of euchromatic and therochromatic zones and a more compact nucleolus (D Chiatante, unpublished).

These few investigations suggest that water might play a crucial role in the structural organization of the chromatin by changing the particular state of dispersion of all its elements. However, the variation of content of nuclear proteins with a structural function or a double structural-enzymic function (DNA topoisomerase II), might also play an active role.

The function of the nuclear membrane in the structural organization of the chromatin inside the nucleus has acquired enormous importance in recent investigations. However, nothing is known about its function during the transition from quiescence to proliferation.

3. Variations of the content of QP47, a novel nuclear protein during the quiescence-proliferation transition

Investigations conducted in my lab have demonstrated that the quiescence-proliferation transition in root meristems is characterized by a considerable variation of protein content in the nuclei [15]. In fact, the attainment of a threshold amount of nuclear proteins might be necessary for the resumption of DNA synthesis [16]. This work has revealed the presence of a nuclear protein which we call "QP47" and which, unlike all the other proteins, seems to be "down regulated" during the quiescence-proliferation transition [4]. Some of the properties of this protein are known. For example, we have determined a partial amino acid sequence, its iso-electric point and acidic nature (D Chiatante, unpublished), but its identity is still largely unknown. A strong correlation exists between the decrease of the amount of QP47 and the transition of root meristems of germinating seeds from quiescence to the state of proliferation. In fact, when this transition is delayed or anticipated we observe a delay or an anticipation of the decrease of QP47 [12].

Also of interest is the observation that phosphorylation of this protein increases as the level of the protein starts to decrease [5].

Recently, a purified preparation of this protein has been used to raise polyclonal antibodies serving to localize its position inside the nucleus. The first investigations carried out with this antibody suggested that this protein was located on the surface of the chromosomes. Moreover, the fluorescence of QP47 seems to follow the profiles of the chromosomes (Fig. 1). An external position of QP47 in relation to all the other components of the chromosomes might very well explain the fact that this protein is solubilized by a concentration of NaCl (0.14 M) so low as to leave intact the structure of the nucleosome.

What is the role of QP47 and how important is its degradation for the release

78

Fig. 1. Purified nuclei have been obtained from root tips of seeds germinated for 10 h. The nuclei have been incubated with QP47 antibody then a secondary antibody labelled with Texas-Red fluorescence has been used. The figure shows that the fluorescence of QP47 is arranged to form filaments which may correspond to the DNA filaments.

of cells from their quiescence? Before trying to answer these questions, it is necessary to consider the similarities between QP47 and H1 histone.

The first similarity is that H1 histone also decreases in amount when the seed undergoes the transition from quiescence to germination [7]. The behaviour of H1 (the linker histone) is unique among the histones because none of the others undergo any variations of content during this transition. Furthermore, the phosphorylation turnover of H1 during this event is similar to that of QP47 [5]; in fact, H1 also becomes phosphorylated when its amount starts to decrease [13]. Another similarity may lie in the location of these two proteins on the chromosomes; H1, owing to its linker function, is external to the nucleosome structure.

In the case of H1, the increase of its phosphorylation state may be necessary for the displacement of this protein from the chromatin, with the consequent decrease of chromatin compactness [14, 17]. The fact that the decrease in the amount of H1 more or less coincides with the onset of DNA transcription and replication suggests that the reduced compactness of the chromatin is necessary to allow the DNA molecule to become accessible to the transcription and replication complexes [9, 10].

Notably, variation of content of histones, which regulate both the accessibility of the chromatin and the replication of DNA, also occurs during the cell cycle [13] . The amount of QP47, however, drops to an undetectable level in nuclei of proliferating meristematic cells and in differentiated cells. Therefore, it is not unreasonable to suggest that its role might be correlated exclusively with the particular state of the chromatin during the quiescence of the dried seed. Perhaps, the chromatin of dehydrated tissues needs a greater compactness during the quiescent period [1], and this condition is achieved only by means of a coincidental high content of H1 and QP47. In fact, in animals factors other than H1 might be involved in chromatin condensation [13].

The presence of QP47 in other higher plants during the quiescence-proliferation transition has been investigated using the polyclonal antibody raised against this protein (Fig. 2). The results show that QP47 is conserved in other members of the Leguminosae family (*Vicia faba* and *Cicer arientinum*) but is missing from the monocots species tested so far (*Zea mays* and *Hordeum vulgare*). In the Leguminosae the correlation between the decrease of the level of QP47 and the quiescence-proliferation transition seems to be confirmed. In the monocot plants, a protein different (in molecular weight, and is not recognized by the QP47 antibody) from QP47 was detected which showed the same decrease in amount coinciding with the release from quiescence (D Chiatante, unpublished). Although more information is necessary, the data suggest that, even if QP47 is not highly conserved in higher plants, its function could be carried out by a different set of proteins in different taxonomic groups.

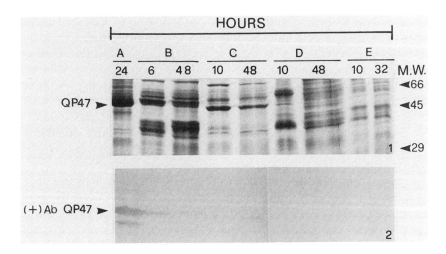

Fig. 2. Electrophoretic patterns obtained with nuclear proteins of root meristems at different hours of germination. A = *Pisum sativum*; B = *Vicia faba*; C = *Cicer arietinum*; D = *Zea mays*; E = *Hordeum vulgare*. 1) On the right the arrows indicate the position of the molecular weight standards. On the left, the arrow shows the position of QP47 protein. 2) The blotting shows the reaction of QP47 antibody with the nuclear proteins. The strongest reaction of QP47 antibody is obtained with pea nuclear protein; nevertheless, some reaction is evident with the other leguminous plants whereas it is absent from the monocots.

4. Activities of enzymes of DNA metabolism during the quiescence-proliferation transition

After the hydration of the nuclei and the changes in the structural organization of the chromatin, the release from quiescence of the cells still cannot be completed without the activation of the enzymes involved in DNA metabolism. During seed maturation, the chromatin undergoes a considerable dehydration, which causes breakage in the DNA molecule. Such breakages must be repaired before any metabolic activity of the DNA can be re-established and the cells successfully released from their quiescence [18].

One of the enzymes involved in DNA repair seems to be DNA topoisomerase I and the activity of this enzyme should increase during the first stage of germination (see above). In roots of *Pisum sativum,* during the first 72 h of germination (in which the quiescence-proliferation transition takes place), the specific activity of this enzyme increases by means of an increase not of the quantity of the protein, but of its phosphorylation state (D Chiatante, unpublished).

The antibodies raised against a DNA topoisomerase I from a human cell antigen, have recognized , in nuclear extracts obtained from root meristematic tissue of *Pisum sativum*, a protein with a molecular weight of 45 kDa. The specificity of this antibody reaction was demonstrated by an immuno-precipitation experiment in which the elimination of a part of this protein resulted in a decrease of DNA topoisomerase activity in the extracts (D Chiatante, unpublished). The amount of the protein recognized by the antibody did not change during the quiescence-proliferation transition.

The DNA polymerases, also, seem to undergo some activation during the first stage of germination, but in this case the mechanism of activation is not clear [3]. In *Pisum sativum*, a DNA polymerase activity with some characteristic of the β-type enzyme may be greater than DNA polymerase activity of the α-type, at least during the first few hours of germination [3]. DNA polymerase β seems to be involved in DNA repair and this would explain the early activation of this enzyme in relation to the other enzymes involved in DNA replication.

The role of proliferating cell nuclear antigen (PCNA) as an activator of DNA polymerase δ is well established. In our investigations, a PCNA-like protein increases in amount during the transition from quiescence to cell proliferation [6]. However, in this case further research is necessary in order to ascertain whether the protein recognized by the antibody is in fact PCNA. If it is, it will be a further type of DNA polymerase subject to a mechanism of activation during the period of transition of cells from quiescence to cell proliferation.

The case of DNA topoisomerase II is special owing to the double action of this enzyme which has a structural role in the chromosome scaffold and also introduces a topological modification to the DNA molecule (see SJ Aves and JA Bryant, this volume). DNA topoisomerase II seems to be more active in highly active meristematic tissues, such as cauliflower inflorescences [8], but nothing is known about the activity of this enzyme during the first phases of germination. We lack a clear demonstration that DNA topoisomerase II is present in pea nuclear preparations owing to the lack of a specific assay serving to exclude the interference of the DNA topoisomerase I present in the same extract. For these reasons, the investigations regarding this enzyme are based, at present, on studies of immunocytofluorimetry (work in preparation).

In *Pisum sativum*, a nuclear protein is recognized by an antibody prepared against a human DNA topoisomerase II, and such immuno-cytofluorimetric investigations suggest that this protein may undergo qualitative variations during the quiescence-proliferation transition. An interpretation of these variations based only on immunological investigations is rather difficult, since unfortunately the same antibody fails to recognize a single protein from the same nuclei in blotting experiments.

5. Conclusions

The release from quiescence of cells in meristematic tissues during the first phases of germination is obviously the result of numerous interdependent physiological events each necessary to ensure the success of this transition. Much physiological work has been done in an effort to understand the factors that influence this transition, whereas the molecular and cytological aspects of this phase of germination have tended to be overlooked. For example while it is known that the onset of DNA replication is a late event in the release from quiescence, there is not the slightest indication of what occurs in the nucleus to prepare the chromatin for its normal activities.

In this paper, some of the events taking place in the nucleus during this transition have been considered from a biochemical and cytological point of view. However, the data presented here, and the literature in general, by no means reveal the entire sequence of these events.

The basic question of whether replication sites are dynamic (move freely inside the nucleus) or fixed on the nuclear matrix, with the DNA "spooling" through them has not yet been answered. I believe that the answer to this question is of primary importance for understanding the mechanisms involved in the release from quiescence. In fact, in the case of replication sites fixed on the nuclear matrix, it is clear that the variations of structural organization of the chromatin occurring in the nucleus during the quiescence- proliferation transition, must be completed before the enzymes forming the replication or transcription complexes are in a condition to start their activities.

This would suggest that a pronounced structural rearrangement of the chromatin (by water hydration and protein rearrangement) in the nucleus is the first stage, preceding the onset of any metabolic reactions involving the DNA molecule. However, even if replication sites are of the dynamic type, their assembly and activation cannot take place until the corresponding binding sites on the DNA molecule have been made accessible by means of a previous variation of structural organization of the chromatin or the elimination of blocking factors (proteins) (see SJ Aves and JA Bryant, this volume).

In both cases, the modification of the structural organization of the chromatin is very important for the successful release of the cell from the quiescence; moreover, in both cases the functional role of QP47 might be crucial by allowing, through its degradation, the increase of accessibity of the DNA molecule. The accuracy of this hypothesis should be proved in future by experiments aimed to investigate the behaviour of this protein during the period of maturation of the seed, when quiescence is enforced on embryo tissues. The rationale is that if QP47 has a role in protecting the DNA molecule during dehydration by increasing its compactness, we should observe an increase of its content in cells during seed maturation. However, there must be other factors controlling the release from quiescence after the action of QP47. A factor could be the activation of the enzymic machinery necessary for DNA replication; another important factor could be the assembly of these enzymes to form the multiprotein complexes active at the replication sites.

The identification of proteins with a functional role in the transition quiescence-proliferation might help to understand the mechanisms of gene regulation of this important physiological event of germination.

Acknowledgements

I thank the National Research Council of Italy, Special Project Raisa, Sub-project N. 2, Paper N 557, for supporting this work.

References

1. Baluska F (1990) Nuclear size, DNA content, and chromatin condensation are different in individual tissues of the maize root apex. Protoplasma 158: 45–52.
2. Bouvier-Durand M, Real M and Come D (1989) Changes in nuclear activity upon secondary dormancy induction by abscisic acid in apple embryo. Plant Physiol Biochem 27: 511–518.
3. Bryant JA, Jenns SM, and Francis D (1980) DNA polymerase activity and DNA synthesis in roots of pea (*Pisum sativum*) seedlings. Phytochemistry 20: 13–15.
4. Chiatante D, Brusa P, Levi M and Sparvoli E (1991) Nuclear proteins during the onset of cell proliferation in pea root meristems. J Exp Bot 42: 45–50.
5. Chiatante D, Brusa P, Levi M and Sparvoli E (1991) Phosphorylation of nuclear proteins and proliferation in root meristem of pea (*Pisum sativum*). Plant Science 75: 39–46.
6. Colombo B, Chiatante D, Citterio S, and Sparvoli E (1992) Biochemical investigation of PCNA (Proliferating cell nuclear antigen) in meristematic cells of *Pisum sativum* during the activation of cell proliferation. Life Science Advances, (in press).
7. Deltour R (1985) Nuclear activation during early germination of the higher plant embryo. J Cell Sci 75: 43–83.
8. Fukata H and Fukasawa H (1986) Isolation and characterization of DNA topoisomerase II from cauliflower inflorescences. Plant Mol Biol 6: 137–144.
9. Garrard WT (1991) Histone H1 and the conformation of transcriptionally active chromatin. Bioassay 13: 87–88.

10. Ivanov PV and Zlatanova JS (1989) Quantitative changes in histone content of the cytoplasm and the nucleus of germinating maize embryo cells. Plant Physiol Biochem 27: 925–930.

11. Krachmarov C, Stoilov L, Zlatanova J (1991) Nuclear matrices from transcriptionally active and inactive plant cells. Plant Sci 76: 35–41.

12. Levi M, Pasini E, Brusa P, Chiatante D, Sgorbati S, and Sparvoli E (1991) Culture of pea embryo axes for studies on the reactivation of the cell cycle at germination. In vitro Cell Dev Biol 28P: 20–24.

13. Osley MA (1991) The regulation of histone synthesis in the cell cycle. Ann Rev Biochem 60: 827–861.

14. Roth SY and Allis CD (1992) Chromatin condensation: does histone H1 dephosphorylation play a role? TIBS 3: 93–98.

15. Sgorbati S, Sparvoli E, Levi M, Chiatante D and Giordano P (1988) Bivariate cytofluorimetric analysis of DNA and nuclear protein content in plant tissue. Protoplasma 144: 180–184.

16. Sgorbati S, Sparvoli E, Levi M, and Chiatante D (1989) Bivariate cytofluorimetric analysis of nuclear protein and DNA relative to cell kinetics during germination of *Pisum sativum* seed. Physiol Plant 75: 479–484.

17. Travers AA (1992) DNA conformation and configuration in protein DNA complexes. Curr Opin Struct Biol 2: 71–77.

18. Zlatanova JS, Ivanov PV, Stoilov LM, Chimshirova KV and Stanchev BS (1987) DNA repair precedes replicative synthesis during early germination of maize. Plant Mol Biol 10 : 139–144.

7. Histone H2A expression during S-phase: histological co-localization of H2A mRNA and DNA synthesis in pea root tips

EUGENE Y. TANIMOTO, THOMAS L. ROST and LUCA COMAI

Abstract

Histone H2A mRNA is expressed developmentally in specific cells in the pea root apical meristem. A double labelling technique was used to identify cells replicating DNA and also expressing H2A mRNA. We found that approximately 90% of the S-phase cells were expressing H2A mRNA. About 10% of the cycling cells expressed H2A mRNA outside of S-phase and about 10% of S-phase cells did not express H2A mRNA. When DNA synthesis was inhibited with hydroxyurea, a commensurate and specific decrease in steady state levels of H2A mRNA was found. We conclude that cell specific expression of pea histone H2A mRNA is dependent on cell cycle regulation and that H2A mRNA is transiently accumulated during a period of the cell cycle which mostly overlaps the S-phase. An assumption that H2A mRNA accumulation was implemented in late G1, and abated in late S could account for this observation.

1. Introduction

Replication of DNA requires a coordinate and proportional synthesis of other components of chromatin [see SJ Aves and JA Bryant, this volume]. Many genes coding for replication dependent products are expressed transiently in an ordered fashion during the cell cycle. Histone proteins are required for packaging of eukaryotic DNA in chromatin and most histone gene expression in metazoans and yeast is dependent on DNA synthesis [1, 8, 15]. As a result of packaging requirements, the regulation of histone gene expression is believed to be fundamental to the control of cell proliferation [16]. For cell cycle-regulated histone genes, transcription is activated near the G1-S transition and mRNAs and proteins accumulate in S-phase [8, 17, 23, 24]. Upstream and downstream elements of the histone genes are responsible for cell cycle control [14]. In contrast, low constitutive expression characterizes cell cycle independent histone genes whose expression may be linked to such things as histone turnover outside of the S-phase [3, 9, 28]. Recently a histone H4 gene has been cloned from *Arabidopsis thaliana* that showed replication dependent and independent expression when monitored as a chimeric H4 promoter-β glucuronidase fusion in transgenic tobacco plants [12].

We have previously reported that tomato H2A.1 and pea histone H2A are

85

J.C. Ormrod and D. Francis (eds.), Molecular and Cell Biology of the Plant Cell Cycle, 85–95.

differentially expressed in developing primary plant tissues [10]. In pea, RNA gel blot analysis and *in situ* RNA hybridization revealed that the highest expression, in terms of the percentage of mRNA or in cell number, occurred within the root apical meristem. Expression then declined basipetally through the primary root. Moderate H2A mRNA levels could be found in the whole shoot and very low levels could be found in mature leaves. Pea roots were chosen to study because they have large well characterized primary roots [26, 19, 21].

A novel combination of two differently detectable labels was used to co-localize cells in S-phase and those cells that were expressing H2A mRNA. S-phase cells were pulse-labelled with ^{3}H-thymidine and detected by autoradiography. H2A mRNA was labelled by *in situ* RNA hybridization with digoxigenin-tagged antisense H2A RNA and detected by the reaction of alkaline phosphatase conjugated to antibodies that were immunoreactive to digoxigenin [2]. The pattern of expression of H2A mRNA with respect to the pattern of proliferation of cells within the apical meristem will be discussed.

2. Materials and methods

Pea seedlings, *Pisum sativum* cv. Alaska, were surface sterilized in 50% bleach (v/v) and 0.75% Alconox® detergent (w/v), rinsed in sterile water and sown in sterile moistened vermiculite for 4 to 5 days in the dark at 22-25°C. Root tips (1 cm long) were isolated from 4 to 7 cm long primary seedling roots.

Unless noted, all solutions used in protocols for nucleic acid manipulation were taken from Sambrook *et al.* [22].

2.1. Culturing of root tips for nucleic acid labelling and extraction

Ten root tips were placed in 125 ml flasks with 50 ml of White's medium [4] pH 5.0 (adjusted prior to autoclaving) plus 2% sucrose. Cultured root tips were pulse-labelled after 48 h of culture on an orbital shaker (60 rpm) at 23°C [27]. Roots were pulse-labelled for 1 h in the same culture medium; experiments were repeated with samples using similar concentrations of isotope ranging from 1 to 5 μCi/ml (methyl-^{3}H-thymidine @ 78.5 Ci/mmol, NEN/Dupont). Root tips were rinsed with H_2O and frozen in liquid N_2 and stored at -80°C.

2.2. Hydroxyurea

Cycling plant cells can be arrested in S-phase with hydroxyurea (HU) treatment [7]. Root tips were grown as before in White's medium plus 2% sucrose and equilibrated on a rotary shaker in the dark for 48 h prior to HU treatment. Root tips were grown for 24 h in the presence of 70 mM HU and were labelled for 1 h with methyl ^{3}H-thymidine as before. Root tips were rinsed with H_2O and frozen in liquid N_2 and stored at -80°C.

2.3. Extraction of nucleic acids

Labelled DNA was extracted by a modified Schmidt-Tannhauser procedure [25]. Frozen roots were quickly homogenized in 0.3 M potassium hydroxide in a ground glass homogenizer. The homogenate was hydrolyzed at 37°C for 1 h and centrifuged at 1500 × g for 10 min. The supernatant was collected and the pH was adjusted to 1.5 with hydrochloric acid then kept on ice for 1 h to allow the DNA to precipitate. Samples were centrifuged at 10,000 × g for 10 min and the pellet washed with cold acidified water. The pellet was heated in an appropriate volume of 0.5 M perchloric acid for 15 min at 80°C. For DNA synthesis measurements, an aliquot of the extract was placed in Aquasol® (Dupont) scintillation fluid and counted with a Beckman scintillation counter. Total RNA was miniprepped from cultured root tips by hot phenol extraction and lithium chloride (LiCl) precipitation as modified from Martineau et al. [13]. Ten roots tips (0.3 g FW) were homogenized in a Polytron® in 2 ml of H-buffer (100 mM Tris-HCl pH 9, 100 mM NaCl, 1 mM EDTA, 0.5% SDS). The homogenate was extracted with 1.5 ml hot H-buffer-equilibrated phenol (65°C) on a shaker for 15 min. An equal volume of chloroform:isoamyl alcohol (24:1 v/v) was added and the samples were vortexed and extracted for 15 additional min. The aqueous phase was separated by centrifugation (10,000 × g for 15 min), then collected and adjusted to 0.2 M with potassium acetate, and precipitated with 2.5 volumes of absolute ethanol at −20°C overnight (or 1 h at −80°C). The pellet was collected and resuspended in Tris-EDTA buffer (TE) and re-precipitated as before. After centrifugation, the pellet was resuspended in TE and adjusted to 2.0 M LiCl by the addition of an equal volume of 4.0 M LiCl. Total RNA was recovered by centrifugation (12,000 × g) after overnight precipitation at 4°C. Usually, 50 to 100 μg of total RNA was isolated from the starting material. RNA gel blots were made according to the protocol described by Fourney et al. [5] with 20 mg of total RNA/lane and transferred to Nytran® (Schleicher and Schuell manufacturer). Quantitation of mRNA expression was determined by densitometry of the exposed image of histone H2A mRNA on the autoradiograph.

Random primed or nick translated probes were synthesized from agarose gel isolated inserts of EcoR1-Xbal digestions of the pGEM 1-H2a cDNA plasmid. Hybridizations were carried out in a solution containing 50% formamide, 1 M NaCl, 1% SDS and 100 mg/ml sheared salmon sperm DNA and final post-hybridization washes were in 0.2 × SSC (standard sodium chloride) [22] and 0.1% SDS at 68°C. Sequencing was done by the dideoxy chain termination method (U.S. Biochemical Corp., Cleveland, Ohio) Sequenase® , 7-deaza sequencing kits.

2.4. Fixation, embedding and sectioning

Roots were grown as above. One-cm root tips were pulse-labelled as before either immediately after excision or after 24 h of culture. Roots were pulse-labelled for 1 h in methyl ^{3}H-thymidine, 1 μCi/ml, (78.5 Ci/mmole, NEN/Dupont) in White's medium plus 2% sucrose, rinsed with water and fixed in 4% paraformaldehyde in phosphate buffered saline (10 mM Na-phosphate, 150 mM NaCl, 5 mM MgCl$_2$, pH 7.4) for 12 h, dehydrated through a standard tertiary butyl alcohol series and embedded in Paraplast® as in Pokalsky *et al.* [18]. Ten 1 mm thick sections were cut with a rotary microtome.

Slides were cleaned in acidified ethanol (95 ml of 95% ethanol, 5 ml of glacial acetic acid), rinsed in deionized water and air dried. Clean slides were coated with polylysine by dipping in 50 mg/ml poly D or L lysine (Sigma) in 10 mM Tris-HCl pH 8.0 for 30 min and air dried. Serial Paraplast® sections, cut to fit under an untreated 22 × 40 mm coverslip, were overlaid on H$_2$O at 45°C on a slide warming tray until the ribbons were expanded. Excess H$_2$O was drawn off with a Kimwipe® to prevent bubbles from forming under the tissue and the slides were dried overnight at 45°C.

2.5. Autoradiography

The slides were de-paraffinized in xylene, dipped in 1:1 (xylene:100% ethanol), then 100% ethanol, 5 min each, and air dried. Slides were coated in diluted Kodak NTB-2 film emulsion (1:1 with H$_2$O), dried and stored at 4°C in an air/light tight box for 1 week. The autoradiographs were developed for 5 min in Kodak D-19 developer, rinsed briefly in H$_2$O, and fixed for 10 min in Kodak general purpose fixer at 15°C. The autoradiographs were washed in diethylpyrocarbonate (DEPC) treated H$_2$O for 10 min. DEPC-H$_2$O was made by adding DEPC to 0.05% (v/v) and mixing thoroughly prior to autoclaving. All darkroom manipulations were done under a 15 W bulb and Kodak 2 red or Kodak GBX-2 filter.

2.6. Probe preparation

Pea cDNA was synthesized from poly (A) RNA isolated from whole 1.5 cm root tips and was cloned into LambdaGem-4 (Promega). Pea histone H2A was isolated from a pea root cDNA library by use of a heterologous tomato H2A cDNA probe [10]. A 627 bp pea H2A cDNA was cloned into pGEM-1 as an *Eco*R1-Xbal fragment (GenBank accession M64838). For *in situ* RNA hybridizations, the plasmids were linearized with EcoR1 or Xbal digestion yielding the sense or antisense template for *in vitro* transcription from T7 or SP6 promoters. Digoxigenin (DIG)-labelled RNA probes were synthesized from 1 μg of linearized plasmid using a Boehringer Mannheim RNA labelling kit. The detectable activity of the probe was tested on dot blots of graded concentrations of double stranded H2A cDNA. Activity was measured using 10% of the synthesized RNA probe. Using a Boehringer Mannheim non-radioactive nucleic acid detection kit, a minimum of 10 pg of H2A cDNA was detectable by usable probe. Normally, transcripts were resuspended in 50 ml of DEPC treated water plus 50 units of RNasin® , Promega, placental ribonuclease inhibitor and stored at $-80°C$. Dot blot detection in 5 ml of hybridization fluid was done according to manufacturer specification with 5 μl of the resuspended probe. For *in situ* hybridization, 1 ml/slide of the resuspended probe was used. No hydrolysis of the transcript was necessary to improve *in situ* hybridization. Probes were stable in frozen aqueous suspensions at $-80°C$ for several months.

2.7. Prehybridization of root sections

The developed autoradiographs were dipped in 100 mM triethanolamine in DEPC-H$_2$O at room temperature. The slides were removed and acetic anhydride was added to make a 0.5% (v/v) solution and the slides were re-immersed for 10 min. The slides were rinsed twice in 2X SSPE (standard sodium phosphate-EDTA) [22], 5 min each, and rinsed twice in DEPC-H$_2$O, 2 min each as in Raikhel *et al.* [20]. Prehybridization consisted solely of this acetylation step in order to block non-specific binding of negatively-charged nucleic acid probe *in situ* [6]. Unlike conventional *in situ* RNA hybridization, no enhancement of hybridization by deproteinization with HCl or proteinase K was done since it was determined that NTB-2 nuclear track film emulsion was made with gelatin.

2.8. Hybridization

Fifty ml of hybridization solution (hyb-sol) per slide was pipetted onto the slide over the wet developed film emulsion. A 22 $\times$ 40 mm coverslip was gently lowered over the hyb-sol and rubber cement was applied in ring around the edges of the coverslip. Cement was applied through a disposable 10 ml syringe and 18 gauge needle. The slides were placed overnight at 60°C in a humid incubator (humidity provided by an open container of water).

Hyb-sol was prepared in the following manner. One ml/slide of the resuspended probe was diluted in 50% formamide and 10 mM dithiothriotol (DTT), heated to 70°C, and mixed with hybridization buffer to make the hyb-sol. Final concentrations of components in the hyb-sol were: 0.3 M NaCl, 0.01 M Tris-HCl (pH .8), 0.01 M Na$_2$-phosphate (pH 6.8), 5 mM Na$_2$-EDTA, 50% formamide, 10% dextran sulfate, 1.0 mg/ml yeast tRNA, 8 mM DTT, 0.5 mg/ml poly (A) and probe. Hybridization buffer and hyb-sol preparation can be found in Raikhel *et al.* [20]. We elected to use DTT to reduce potential background; normally it is used to prevent ^{35}S-labelled probes from forming disulfide bridges.

2.9. Post hybridization

After hybridization, all solutions can be made with non-treated water. The rubber cement was removed with forceps, being careful not to peel off the cover-slip. The slides were soaked with gentle agitation in 2 $\times$ SSC until the coverslips floated off and then the slides were washed twice in 2 × SSC for 1 h each; once in 1 × SSC for 1 h, and once in 0.5 × SSC for 1 h all at room temperature.

2.10. Immunological detection[1]

The slides were washed in Buffer # 1 for 5 min at room temp., then blocked by complete immersion in 2% normal sheep serum (NSS,Cal-Biochem), 0.3% Triton X-100 in Buffer # 1 for 30 min at room temperature. Slides were taken out of the blocking solution, and 200 ml/slide of diluted antibody (diluted 1:500 in Buffer # 1 containing 1% NSS and 0.3% Triton X-100) was applied for 2 h at room temp. The slides were washed twice in Buffer # 1 for 15 min at room temp., then washed once in Buffer # 3 for 2 min at room temp. For detection, 500 ml of freshly made colour solution (nitroblue tetrazolium salt and X-phosphate) made in Buffer # 3 was applied to each slide and incubated in a humid chamber in the dark at room temp for 2 to 3 h. The colour reaction was stopped in Buffer # 4 for 5 min. The slides were dehydrated through an ethanol series, passed through xylene and cover slips were mounted with Permount® . Appreciable colour precipitate was lost in an extended stay in the xylene:ethanol (1:1) step. If omitted, fresh 100% ethanol should be used to prevent clouding of sections, due to water in the tissue, when passed into xylene.

3. Results

3.1. Co-localization of histone H2A mRNA with cells in S-phase

To determine the association of histone H2A mRNA expression to the cell cycle, a novel combination of two procedures was developed. S-phase cells were labelled with ³H-thymidine and detected by autoradiography. H2A mRNA was localized by in situ RNA hybridization with digoxigenin (DIG) labelled antisense RNA and immunolocalized with the Genius® non-radioactive Nucleic Acid Detection kit. Figure 1 shows that H2A mRNA was expressed in cells distributed throughout the root tip. The highest intensity of silver staining pertaining to H2A mRNA was in the root apical meristem, and it declined basipetally through the primary root until it ceased in mature cells except in the xylem pericycle cells located 10 to 20 mm from the apex. Expression of H2A mRNA and S-phase cells are noticeably absent in cells of the quiescent centre and mature root cap (Fig. 1).

Cells expressing histone H2A and undergoing DNA synthesis did not overlap

[1] These manipulations followed the recommendations in the Boehringer Mannheim Technical update 8812395/2 m (1989) for the Genius® Nonradioactive DNA Labeling and Detection Kit. Solutions: Buffer 1: 100 mM Tris-HCl, 150 mM NaCl (pH 7.5), Buffer 3: 100 mM Tris-HCl, 100 mM NaCl, 50 mM MgCl₂ (pH 9.5), Buffer 4: 10 mM Tris-HCl, 1 mM EDTA (pH 8.0).

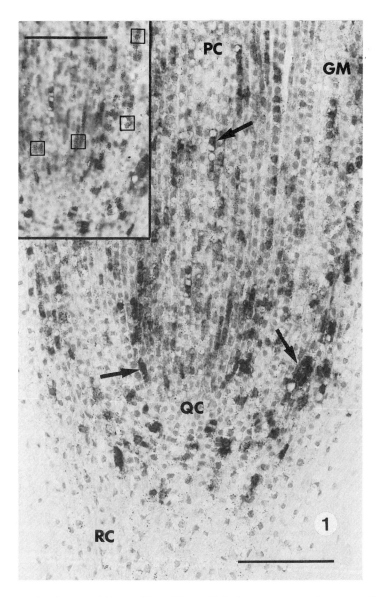

Fig. 1. Co-localization of histone H2A mRNA with S-phase cells in a median longitudinal section of a pea root apical meristem. Detection of H2A mRNA hybridized, *in situ*, to DIG-labelled anti-sense H2A RNA: Most cells containing H2A mRNA (dark stained cells, arrows show three examples) were co-localized with S-phase cells (silver grains above nuclei, compare boxed cells in insert at upper left of a similar section focused at the level of the emulsion). Co-localization was found in all primary meristems within the root apical meristem, procambium (PC), ground meristem (GM), and protoderm. No labelling of S-phase or H2A mRNA expressing cells were found in the quiescent centre (QC) or mature root cap (RC). Line scales = 100 μm.

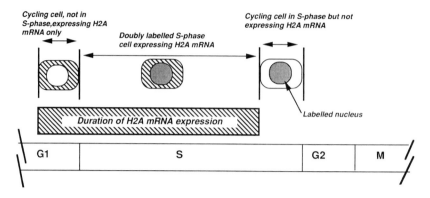

Fig. 2. Timing of H2A mRNA expression during cell cycle. The period of H2A mRNA expression during the cell cycle is shown at the hatched bar. Singly labelled cells expressing DIG-labelled H2A mRNA are detectable in late G1 (dark stained cells in Fig. 1, are represented by diagonal striped cells). Singly labelled S-phase cells (contain ³H-thymidine, but no DIG-label) are detectable in late S. Doubly labelled cells are seen when S-phase and H2A mRNA expression overlap.

exactly. About 90% of the cells in S-phase were found to be expressing H2A mRNA (these were double labelled). Approximately 10% of the S-phase cells were not expressing H2A mRNA (silver grains only) and about 10% of cells expressing H2A mRNA were not in S-phase (DIG label only). We eliminated the possibility that incomplete overlap was caused by differences in the plane of section by counting only entire cells. Because of the packaging requirements of DNA into chromatin, we assumed that replication of DNA should not begin until histone proteins were present. This meant that H2A mRNA would need to accumulate prior to S-phase. Figure 2 demonstrates one way to account for the incomplete overlap through the timing of H2A mRNA accumulation and degradation during the cell cycle.

Similar co-localization of H2A mRNA expression in S-phase cells occurred in cultured roots 24 to 48 h post-excision (data not shown), but fewer labelled cells were found in the root apex.

3.2. Inhibition of DNA synthesis and H2A mRNA expression

The dependency of expression of H2A mRNA on DNA synthesis was investigated. DNA synthesis was arrested by hydroxyurea (HU), a known inhibitor of ribonucleoside diphosphate reductase [7, 11]. Quantitation of the inhibition was estimated by the amount of ³H-thymidine incorporated into DNA. RNA gel blot analysis of H2A mRNA and a constitutively expressed gene from pea roots, PC47, was compared. PC47 mRNA is highly expressed in all organs of the pea plant, including all tissues of the primary root including the root cap. HU inhibited roots synthesized 10% as much DNA as untreated roots. Histone H2A mRNA expression was also reduced by 90% compared to the untreated controls while the constitutively expressed PC47 mRNA was reduced by only 18% after HU treatment.

4. Discussion

Using a novel combination of two detection methods, we have shown that histone H2A mRNA was expressed differentially within the root apical meristem in a cell specific manner that was dependent on cell cycle regulation. H2A mRNA was expressed during a period of the cell cycle that mostly overlaps the S-phase of meristematic cells; steady state levels of mRNA accumulate in late G1 and decline in late S. The *in situ* findings of this study corroborated circumstantial evidence for cell cycle dependency on expression of H2A mRNAs from complex tissues after cell cycle inhibition by hydroxyurea. To our knowledge, this is the first report that describes the timing of histone expression in complex tissues in organs of higher plants without induced synchronization. This observation is consistent with the observations from single cell cultures of yeast and human HeLa cells that histone transcription starts in late G1 and mRNA and protein accumulate in S-phase [14].

In situ RNA hybridization with non-radioactive digoxigenin labelled anti-sense RNA, provided a higher degree of cell to cell resolution in the detectability of H2A mRNA than previous ^{35}S-antisense RNA labelling [10]. In combination with autoradiographic detection of ^{3}H-labelled S-phase cells, digoxigenin labelling provided less ambiguity in localization of cell specific processes than conventional double isotope-autoradiography techniques which use two layers of film emulsion over the tissue to differentially detect low and high energy β-particles emitted by isotopes such as ^{3}H and [35]S. In developing this novel double label protocol, we were forced to rearrange a more logical progression of detection. Autoradiography, using photographic film emulsion to detect ^{3}H-labelled S-phase cells, would normally have been done after *in situ* RNA hybridization and colorimetric detection of the DIG label. We found, however, that the substrates and/or products for colorimetric detection of alkaline phosphatase chemically exposed the photographic emulsion. Similarly, co-localization did not work if colour detection of the H2A mRNA with chemographic alkaline phosphatase substrates was done after autoradiography where *in situ* RNA hybridization was done prior to the application of film emulsion. In lieu of wholesale changes to our planned method of detection, we elected to do the autoradiography first, and hybridize and colour detect the antisense DIG labelled transcript through the developed film emulsion. The result was a histological image with the least amount of background coloration that we have found with any DIG labelled *in situ* hybridization protocol.

References

1. Baumbach LL, Marashi F, Plumb G, Stein G and Stein J (1984) Inhibition of DNA replication coordinately reduces cellular levels of Core and H1 Histone mRNAs: Requirement for protein synthesis. Biochemistry 23: 1618-1625.
2. Bochenek B and Hirsch AM (1990). *In-situ* Hybridization of nodulin mRNAs in rood nodules

using non-radioactive probes. Plant Mol Biol Rep 8: 237-248.

3. Dalton S, Robins AJ, Harvey RP and Wells JRE (1989) Transcription from the intron-containing chicken histone H2A^F gene is not S-phase regulated. Nucleic Acids Res 17: 1745-1756.

4. Dodds JH and Roberts LW (1982) Experiments in Plant Tissue Culture, pp. 30-31. Cambridge: Cambridge University Press.

5. Fourney RM, Miyakoshi O, Day RS and Paterson MC (1988) Northern blotting: Efficient RNA staining and transfer. Focus 10: 5-7.

6. Hayashi S, Gillam IC, Delaney AD, Tener GM (1978) Acetylation of chromosome squashes of *Drosophila melanogaster* decreases the background in autoradiographs from hybridization with [^{125}I]-labelled RNA. J Histochem Cytochem 26: 677-679.

7. Heinhorst S, Cannon G and Weissbach A (1985) Chloroplast DNA synthesis during the cell cycle in cultured cells of *Nicotiana tabacum*: Inhibition by nalidixic acid and hydroxyurea. Arch Bioch Biophys 239: 475-479.

8. Hereford LM, Osley MA, Ludwig JR and McLaughlin CS (1981) Cell-cycle regulation of yeast histone mRNA. Cell 24: 367-375.

9. Huh NE, Hwang IW, Lim K, You KW and Chae CB (1991) Presence of a bi-directional S phase-specific transcription regulatory element in the promoter shared by testis-specific TH2A and TH2B histone genes. Nucl Acids Res 19: 93-98.

10. Koning AJ, Tanimoto EY, Kiehne K, Rost T and Comai L (1991) Cell-specific expression of plant histone H2A genes. Plant Cell 3: 657-665.

11. Krakoff I, Brown NC and Reichard P (1968) Cancer Res 28: 1559-1565.

12. Lepetit M, Ehling M, Chaubet N and Gigot C (1992) A plant histone gene promoter can direct both replication-dependent and – independent gene expression in transgenic plants. Mol Gen Genet 231: 276-285.

13. Martineau B, McBride KE and Houck CM (1991) Regulation of metallocarboxypeptidase inhibitor gene expression in tomato. Mol Gen Genet 228: 281-286.

14. Moran L, Norris D, Oslet MA (1990) A yeast H2A-H2B promoter can be regulated by changes in histone gene copy number. Genes and Development 4: 752-763.

15. Old RW and Woodland HR (1984) Histone genes: Not so simple after all. Cell 38: 624-626.

16. Owen TA, Holsthuis J, Markose E, Van Wijnen AJ, Wolfe SA, Grimes SR, Lian JB and Stein GS (1990) Modifications of protein-DNA interactions in the proximal promoter of a cell-growth-regulated histone gene during onset and progression of osteoblast differentiation. Proc Natl Acad Sci USA 87: 5129-5133.

17. Plumb M, Stein J and Stein G (1983). Coordinate regulation of multiple histone mRNAs during the cell cycle of HeLa cells. Nucl Acids Res 11: 2391-2410.

18. Pokalsky AR, Hiatt WR, Ridge N, Rasmussen R, Houck CM and Shewmaker, CK (1989) Structure and expression of elongation factor 1a in tomato. Nucl Acids Res 17: 4661-4673.

19. Popham, RA (1955) Levels of tissue differentiation in primary roots of Pisum sativum. Amer J Bot 42: 529-540.

20. Raikhel NV, Bednarek SY and Levner D (1988) *In-situ* hybridization in plant tissues. In: Plant Molecular Biology Manual. SB Gelvin and RA Schilperoot (eds), pp. 1-32, Section B-9. Dordrecht, The Netherlands: Kluwer Academic Publishers.

21. Rost TL, Jones TJ and Falk RH (1988). Distribution and relationship of cell division and maturation events in *Pisum sativum* (Fabaceae) seedling roots. Amer J Bot 75: 1571-1583.

22. Sambrook J, Fritsch EF and Maniatis T (1987) Molecular Cloning: A laboratory manual, 2nd ed. (Cold Spring Harbor, NY: Cold Spring Harbor Laboratory).

23. Schumperli D (1986) Cell-cycle regulation of histone gene expression. Cell 45: 471-472.

24. Shalhoub V, Gerstenfeld LC, Collart D, Lian JB and Stein GS (1989). Down regulation of cell growth and cell cycle regulated genes during chick osteoblast differentiation with the reciprocal expression of histone gene variants. Biochemistry 28: 5318-5322.

25. Tanimoto E, Douglas C and Halperin W (1979) Factors affecting crown gall tumorigenesis in tuber slices of Jerusalem artichoke (*Helianthus tuberosus*, L.). Plant Physiol 63: 989-994.

26. Torrey JG (1955) On the determination of vascular patterns during tissue differentiation in excised pea roots. Amer J Bot 42: 183-198.
27. Van 't Hof J (1965) Cell population kinetics of excised roots of *Pisum sativum*. J Cell Biol 27: 179-189.
28. Wells D and Kedes L (1985) Structure of human histone cDNA evidence that basally expressed histone genes have intervening sequences and encode polyadenylated mRNAs. Proc Natl Acad Sci 82: 2834-2838.

8. Molecular characterization of cell populations in the maize root apex

PAOLO A. SABELLI, SHIRLEY R. BURGESS, JESUS V. CARBAJOSA, JILL S. PARKER, NIGEL G. HALFORD, PETER R. SHEWRY and PETER W. BARLOW

Abstract

Several features make the root apex of maize (*Zea mays* L.) a good experimental system for studying cell cycle controls in relation to development in higher plants. Actively dividing meristematic cells, slowly-cycling quiescent centre cells, differentiating cap cells and senescent detaching cap cells are distinctively compartmented. In addition, the quiescent centre can be activated into rapid proliferation in response to stresses. Although there is a wealth of cytological and physiological information about the behaviour of cells in the root apex, very little is known about the molecular factors which control the patterns of cell division and differentiation. Differential screening of cDNA libraries obtained from discrete cell populations from the apex may provide a means to identify genes which are expressed in cell cycle- and differentiation-dependent manners.

1. Introduction

In vascular plants, apical meristems (the shoot meristem and the root meristem), formed early in embryogeny, are permanently proliferative centres. As such, they are responsible for the production of tissues within the elongating plant body [16, 42]. Secondary meristems of post-embryonic origin (the vascular and cork cambia) also contribute to development with the formation of additional tissue types. In contrast, animals do not have apical meristems. In vertebrates, the organization of the adult body is laid down in the embryo and post-embryonic development consists mostly of the enlargement of embryonic structures [38]. Cell proliferation is then retained mostly in internal cell populations which maintain the size and structure of the adult body.

The activity of the apical meristems leads to the continuous growth, typical of plants. The co-ordinated control of cell division, expansion and differentiation is an essential feature of these apical meristems and is ultimately responsible not only for the development of all the plant structures but also for the accomplishment of vegetative and sexual reproduction.

The root apex represents an excellent system to study the control of cell division in relation to differentiation and development. As will be described in the following sections, cell proliferation and differentiation are regulated in the

J.C. Ormrod and D. Francis (eds.), Molecular and Cell Biology of the Plant Cell Cycle, 97–109.

root tip both in space and time [9]. Complex developmental processes (from quiescence through division and differentiation, to senescence) occur within a few millimeters and within a relatively short period of time. Moreover, in this root system the study is simplified by the absence of the formation of lateral primordia and the transition to an alternative mode of growth (phase change, reproductive development), both of which are events typical of the shoot meristem. Although cytological, histological and physiological analyses have provided a significant amount of information about the physiology of the system, the availability of advanced techniques of molecular biology such as the polymerase chain reaction (PCR) [40] allows for the molecular analysis of factors which control the organization within the root meristem and, ultimately, of the whole root. Here, we describe how differential screening of PCR-amplified cDNA libraries from dissected and discrete cell populations of the maize root apex may prove to be a valuable tool to identify genes involved in cell cycle and differentiation events.

2. The experimental system

Since the early fifties histological analyses of cell lineages and physiological studies coupled with the use of radiolabelled nucleic acid precursors have shown that the root apex consists of structurally and physiologically distinct populations of cycling cells [20]. The maize root tip is an excellent experimental system because it is probably the best characterized, and its relatively large size and structure allow manual dissection. As indicated in Fig. 1, it can be divided, both physiologically and physically, into four regions:
1. The proliferative meristem which can be sub-divided into the meristem proper, sometimes called the "proximal meristem", located just above the quiescent centre, and the cap meristem, located at the cap boundary, which includes the cap initials.
2. The quiescent centre (QC), a small population of approx. 10^3 cells which is located between the proximal meristem and the cap meristem, characterized by low rates of metabolic activity and cell proliferation [14].
3. The root cap which is a mixture of dividing, endoreduplicating and differentiating cells located at the apex tip [5].
4. Senescent cap cells which are non-dividing, terminally differentiated and continuously sloughed from the outer layer of the root cap.
   The striking contrast in the rates of cell proliferation between meristematic cells and QC cells is perhaps the most remarkable feature of the root apex. Earlier analyses based on pulse-labelling with [^{3}H]thymidine [2, 3, 17] have shown that the average duration of the mitotic cycle in the QC is 10- to 15-fold longer than in the meristem (about 170 h and 14 h respectively) and that only approximately 40% of its cells cycle at a rate 3.5- to 4-fold lower than meristematic cells. The majority of these non-dividing cells appear to be in a stationary G1 condition (perhaps better defined as G0) while cycling QC cells

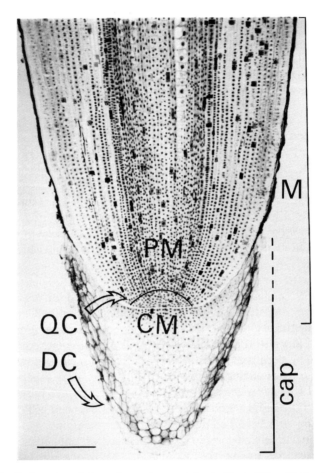

Fig. 1. Longitudinal section of the maize root apex stained with AgNO₃ with the different cell populations indicated. Abbreviations: M, meristem; PM, proximal meristem; CM, cap meristem; QC, quiescent centre DC, detaching cap cells. Bar = 200 μm.

have an approximately 10-fold longer G1-phase. Low levels of signal detected by *in situ* hybridization with histone H1 and H4 anti-sense RNA probes have indirectly confirmed that the rate of DNA synthesis in the QC is remarkably low (SR Burgess *et al.*, unpublished data). This suggests that there must be a block at the G0/G1- to S-phase transition in the QC. Several hypotheses to explain such mitotic inactivity, including the influence of hormones, nutrition, and physical constraints, have been the subject of discussion [4, 19, 43]. However, no conclusive evidence has been presented. In molecular terms, this block can be regarded as being caused by either the absence or inactivation of some factors normally required for the progression of the cell cycle, or by the presence or activation of specific factors which inhibit the cell cycle in the QC.

The comparison of quiescent and meristematic cells could prove an attractive methodological approach to identify specific factors, either activators or inhibitors, of either of the two groups of cells. It is, however, important to emphasize the functional role that may underline the compartmentation of cell division. Although biochemical and physiological properties define the QC as a metabolically inactive cell population [18, 28, 29, 31, 36, 37], several lines of evidence reveal the QC to have an essential role in terms of meristem organization and root development. Soon after its discovery [14], it was unequivocally shown that the QC contains cells that appear to be the origin of the cellular patterns observed in the apex [15]. The QC is also established early in embryogeny and before the formation of histological patterning within the meristem [21, 37]. The formation of a QC before the organization of a root meristem has been also observed in primordia of lateral roots [21] and during apex regeneration following surgical manipulation [4, 25]. Moreover, the size of the QC can affect the histological patterns of neighbouring regions [8, 24] and it is able, following dissection and culture, to regenerate the root (without callus formation) in the same polarity as the root from which it has been isolated [26].

In addition to the slow rate of division, other properties define the QC cells as a stem cell population [6]: they are undifferentiated self-maintaining cells, they produce meristematic and differentiated derivatives, and they can regenerate the tissue or organ with which they are associated if damage occurs. This latter feature is of particular interest. Indeed, the QC can be stimulated into division by a range of stresses, such as damage to the root cap, exposure to cold, exposure to ionizing radiations or refeeding after a period of sucrose starvation [19, 44]. It is generally thought that resistance to these factors lies in the inactivity of the QC. The onset of DNA synthesis and cell division can be monitored by *in vivo* labelling with tritiated thymidine [4], or by *in situ* hybridization with histone probes (S.R. Burgess *et al.* unpublished data). So, in response to external stresses the QC cells are able to overcome the block at the transition from G0/G1- to S-phase and to progress through several cycles of division before re-establishing a new QC. The result is the regeneration of a new cap meristem, a new proximal meristem or both, depending on which cells have been affected by the stress.

Taken together, these data strongly suggest an organizational role of the QC which extends well beyond a mere general metabolic inactivity. The harmonized control of rates of proliferation within the QC and the meristem seems to regulate the organization and activity of the whole apex.

Despite this fundamental role of the QC, very little is known about the factors involved in establishing quiescence or releasing it in response to stress. Similarly, remarkably little is known of the molecular controls of meristem organization and function. Analysis of cell proliferation in the meristem and QC can be coupled to studies on cell differentiation in the root cap. This latter structure, which in maize consists of about 10^4 cells renewed approximately every 1–3 days [7, 17], presents a gradient of differentiation from rapidly dividing cap meristem cells, through endoreduplicating [5] and differentiating cells, to terminally differentiated, senescent cells which are continuously detached from the outermost layer of the root cap (Fig. 1). The root cap is also thought to be the site of perception and transduction of extracellular signals [27]. Thus, in the root apex, integrated regulation of cell proliferation, differentiation and cell-to-cell communication presides over root morphogenesis and development.

The availability of the relatively homogeneous cell populations indicated in Fig. 1 is a valuable asset for biochemical and molecular studies. In addition, at least in maize, individual groups of cells can be dissected from the root apex. This is due to the relatively large size of the root tip, its 'closed organization' with a thick wall at the border between the QC and the cap, and the discrete nature of the QC. The series of complex developmental processes which occur within a very short distance, coupled to our ability to perturb the pattern of development by externally applied stresses and to isolate individual cell populations by manual dissection, provides a system in which to study the molecular controls of plant development (including aspects of cell division, differentiation and senescence). Molecular studies can be based on the analysis of genes expressed in a cell cycle-dependent manner. In the past, differential screening has proved to be a powerful means to identify genes specifically expressed at certain stages of the cell cycle in both animals and plants [22, 33]. Although post-transcriptional modifications are known to be important for the modulation of activity of factors controlling the cell cycle, transcriptional regulation appears to be at least as important [34]. PCR-mediated construction of cDNA libraries makes it feasible to use cloning approaches based on the analysis of differential gene expression in minute portions of tissue [1, 10, 23, 30, 32].

3. PCR-amplified cDNA libraries

In order to study differential gene expression in the root apex, we have constructed PCR-amplified cDNA libraries from the meristem, the QC (both in a normal inactive state and activated into division by removal of the root cap) and the root cap, using cell populations from 10 three-day-old root apices of maize cv. LG11 grown under sterile conditions. Activated quiescent centres were isolated 16 h after removal of the root cap, which corresponds to the time required for activation of DNA synthesis and mitosis (for roots grown at 23 °C) [4]. The cell populations were homogenised by repeated freezing and thawing, together with vortexing in guanidinium chloride buffer containing glass sand. Total RNA was extracted with acid phenol/chloroform [13] to remove any contaminating DNA, and was ethanol precipitated using *E.coli* transfer RNA as carrier. About one third of the unfractionated RNA was reverse-transcribed into cDNA using a commercial kit (Boehringer), using the manufacturer's recommended conditions with minor modifications. Lower concentrations of reverse-transcriptase were used and the RNase inhibitor from human placenta was not included as this may contain contaminating genomic DNA. About 30%

of the double stranded cDNA was then ligated to annealed adaptors and amplified using the longer oligonucleotide in the adaptor as a PCR primer [1]. The conditions for PCR amplification were: denaturation 94°C, 1 min; annealing 60°C, 1.5 min; polymerisation 75°C, 3.5 min, for 30 cycles with 5 sec/cycle extension in the polymerisation step. Standard concentrations of nucleotides, primers, salts, and Taq polymerase (Promega) were used except that a relatively high concentration of Mg^{2+} (3 mM) was found to be optimal

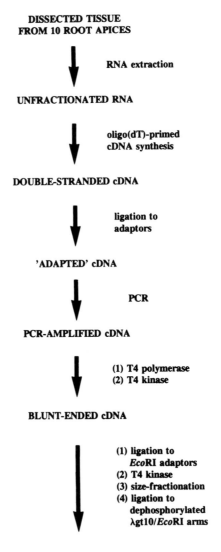

DISSECTED TISSUE
FROM 10 ROOT APICES

RNA extraction

UNFRACTIONATED RNA

oligo(dT)-primed
cDNA synthesis

DOUBLE-STRANDED cDNA

ligation to
adaptors

'ADAPTED' cDNA

PCR

PCR-AMPLIFIED cDNA

(1) T4 polymerase
(2) T4 kinase

BLUNT-ENDED cDNA

(1) ligation to
    EcoRI adaptors
(2) T4 kinase
(3) size-fractionation
(4) ligation to
    dephosphorylated
    λgt10/EcoRI arms

PCR-AMPLIFIED cDNA LIBRARY

Fig. 2. Schematic flow-chart of the procedure followed for the construction of PCR-amplified cDNA libraries.

for the amplification of large sequences (over 4 kb). The amplified cDNAs were blunt-ended using T4 polymerase and phosphorylated by T4 polynucleotide kinase, then ligated to *Eco*RI adaptors, phosphorylated and size-fractionated on Sephacryl S-400 spin columns. cDNAs larger than 400 bp were ligated to dephosphorylated λgt10/*Eco*RI arms and packaged using commercial packaging extracts according to the manufacturer's instructions (Amersham). The protocol is shown schematically in Fig. 2.

4. Identification of cell population-specific cDNAs

An agarose gel separation of the PCR-amplified cDNAs is shown in Fig. 3.

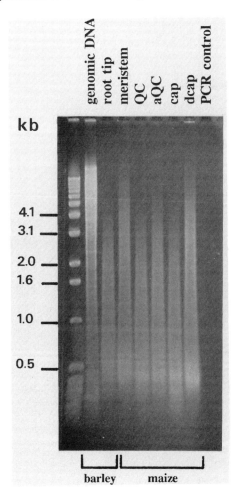

Fig. 3. Agarose gel electrophoresis of cDNAs after PCR amplification. aQC represents activated QC and dcap is detaching cap cells. For the PCR control (far right) no DNA template was added.

This compares the PCR-amplified cDNAs from the different cell populations of maize root tips with a similarly amplified cDNA fraction from whole root tips of barley obtained by preliminary isolation of poly(A)$^+$ RNA, and with barley genomic DNA digested with HindIII. The sizes of the amplified products in all the fractions range from less than 500 bp to over 4 kb. Similar separations were transferred to a nylon membrane and hybridized, under conditions previously described [39], with a histone H3 cDNA from barley [12] (Fig. 4), which belongs to the family of histone genes preferentially expressed during the S-phase of the cell cycle [35]. This probe showed varying intensities of hybridization to the different cDNA fractions, being strongest with that from the meristem and weakest with the fraction from the detached cap cells. *In situ* hybridization of histone H1 and H4 anti-sense RNA probes to sections of maize root tip confirmed this result, except that hybridization in the QC was almost undetectable (results not shown). These data indicate that the relative levels of histone expression in the different cell populations are related to their rates of

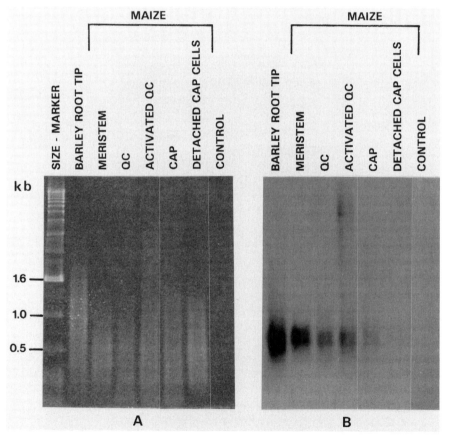

Fig. 4. Southern blot analysis of PCR-amplified cDNAs. A, ethidium bromide staining; B, hybridization to a barley histone H3 cDNA.

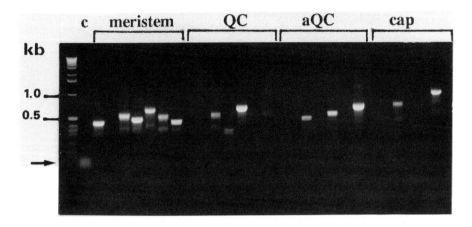

Fig. 5. Agarose gel electrophoresis of randomly PCR amplified bacteriophage λgt10 plaques from cDNA libraries from the meristem, QC, activated QC (aQC) and cap. The arrow indicates the band of PCR primers in the negative control (c).

cell proliferation. The relatively high levels of histone H3 expression in the QC fraction in Fig. 4 is probably due to contamination with meristematic cells. Discrete hybridization bands of the expected size (Fig. 4) confirm that faithful PCR amplification has occurred.

The cDNA libraries contain from 2×10^5 to 5×10^5 clones and PCR amplification of random plaques [40] showed that 40–80% contain amplifiable inserts with an average size of about 700 bp (Fig. 5). The absence of the primers (indicated by an arrow in the control lane) from all samples suggests that some amplification (perhaps not sufficient for ethidium bromide staining) has occurred even in the lanes without any apparent PCR products, and this may indicate a higher library titre.

Differential screening with the different ^{32}P-labelled cDNA probes is currently being carried out to identify cell population-specific clones. The example in Fig. 6 shows some of the putative clones which are preferentially expressed in the root cap and the QC. The screening is being carried out with the aim of identifying genes expressed in a cell cycle-dependent fashion. For example, genes expressed only in the QC may include inhibitors of the cell cycle; those expressed in both the meristem and the activated QC may be putative activators; and those specific to either of these two proliferative cell populations (meristem and activated QC) may be related to the specific proliferative patterns observed. In addition, root cap-specific transcripts could be involved in differentiation as well as in signal transduction [11, 27] (Table 1). Of course, the transcription of most histone genes appears to be cell cycle regulated [35] (see EY Tanimoto *et al.* this volume), but they are outside the scope of this study and could be eliminated from the range of positive clones by screening with a mixture of histone probes.

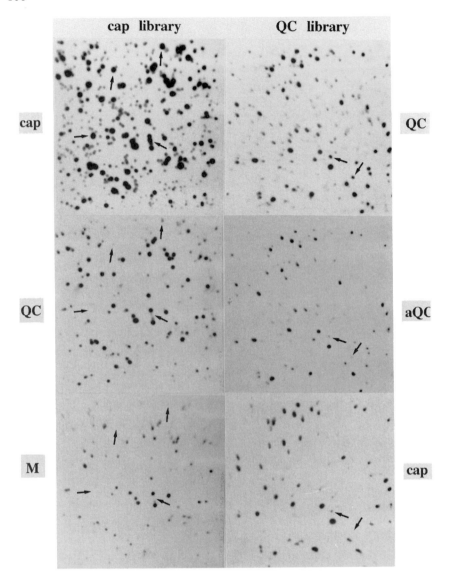

Fig. 6. Example of identification of putative clones specifically expressed in the cap and QC. Aliquots of the cap library (left) and QC library (right) plated at low density have been hybridized to cap and QC cDNA probes, respectively, as positive controls. Subsequent hybridizations of the same (or replica) filters with other cDNA probes as indicated allow the identification of preferentially or specifically expressed clones in the cap and QC libraries (indicated by arrows).

Screening of subtracted cDNA libraries should increase the chances of identifying low-abundant transcripts which are more likely to include regulatory factors [41]. In a preliminary experiment we have prepared QC-specific cDNAs by subtraction with meristematic cDNAs. However, this gave

Table 1. Rationale of the screening procedure. Symbols indicate how, for a given clone, presence of expression (+), absence of expression (−) and possible expression ((+)) may suggest a functional role. M, QC, aQC and cap stand for meristem, quiescent centre, activated quiescent centre and cap libraries, respectively.

Gene expression				Possible role
M	QC	aQC	cap	
−	+	−	−	Cell cycle inhibitor
+	−	+	(+)	Cell cycle activator
−	−	+	−	QC proliferative programme
+	−	−	−	M proliferative programme
(+)	−	(+)	+	Required for differentiation

only a low level of enrichment (about two-fold). This is presumably due to the fact that subtraction experiments in which both the test and the driver sequences are double stranded are less efficient than when the driver sequences are single-stranded molecules of RNA, as these cannot re-anneal (PA Sabelli *et al.*, unpublished).

It is anticipated that cell population-specific clones will be valuable for studying the molecular elements involved in the control of cell division and differentiation, and the tissue-specific control of gene expression, within the root apex.

Acknowledgements

We thank Andy Phillips and Alison Huttly for many helpful discussions and for providing oligonucleotides. This work is supported by the AFRC Plant Molecular Biology Programme (Ref. No. PG 206/517).

References

1. Akowitz A and Manuelidis L (1989) A novel cDNA /PCR strategy for efficient cloning of small amounts of undefined RNA. Gene 81: 295–306.
2. Barlow PW (1973) Mitotic cycles in root meristems. In: The Cell Cycle in Development and Differentiation. M Balls and FS Billet (eds) pp. 133–165. Cambridge University Press, Cambridge.
3. Barlow PW and Macdonald PDM (1973) An analysis of the mitotic cell cycle in the root meristem of *Zea mays*. Proc R Soc B 183: 385–398.
4. Barlow PW (1974) Regeneration of the cap of primary roots of *Zea mays*. New Phytol 73: 937–954.
5. Barlow PW (1977) The time-course of endoreduplication of nuclear DNA in the root cap of *Zea mays*. Cytobiologie 16: 98–105.
6. Barlow PW (1978) The concept of the stem cell in the context of plant growth and development. In: Stem Cells and Tissue Homeostasis. BI Lord, CS Potten and R Cole (eds) pp. 87–113. Cambridge University Press, Cambridge.
7. Barlow PW (1978) Cell displacement through the columella of the root cap of *Zea mays* L. Ann Bot 42: 791–799.

8. Barlow PW and Rathfelder EL (1984) Correlations between the dimensions of different zones of grass root apices, and their implications for morphogenesis and differentiation in roots. Ann Bot 53: 249–260.

9. Barlow PW (1992) The cell division cycle in relation to root organogenesis. Plant Growth Regulator Supplement (in press).

10. Belyavsky A, Vinogradova T and Rajeewsky K (1989) PCR-based cDNA library construction: general cDNA libraries at the level of a few cells. Nucleic Acid Res 17: 2919–2932.

11. Biermann B, Johnson EM and Feldman LJ (1990) Characterization and distribution of a maize cDNA encoding a peptide similar to the catalytic region of second messenger dependent protein kinases. Plant Physiol 94: 1609–1615.

12. Chojecki J (1986) Identification and characterization of a cDNA clone for histone H3 in barley. Carslberg Res Commun 51: 211– 217.

13. Chomczynski P and Sacchi N (1987) Single step method of RNA isolation by acid guanidinium thiocyanate-phenol-chloroform extraction. Anal Biochem 162: 156–159.

14. Clowes FAL (1954) The promeristem and the minimal constructional centre in grass root apices. New Phytol 53: 108–116.

15. Clowes FAL (1956) Nucleic acids in root apical meristems of *Zea*. New Phytol 55: 29–34.

16. Clowes FAL (1961) Apical meristems. Blackwell Scientific Publications, Oxford.

17. Clowes FAL (1971) The proportion of cells that divide in root meristems of *Zea mays* L. Ann Bot 35: 249–261.

18. Clowes FAL (1972) The control of cell proliferation within root meristems. In: The Dynamics of Meristem Cell Populations. MW Miller and CC Kuehnert (eds) pp. 133–147. Plenum Publishing Corporation, New York.

19. Clowes FAL (1975) The quiescent centre. In: The Development and Function of Roots. JG Torrey and DT Clarkson (eds) pp. 3–19. Academic Press, London.

20. Clowes FAL (1976) The root apex. In: Cell Division in Higher Plants. MM Yeoman (ed) pp. 253–284. Academic Press, London.

21. Clowes FAL (1978) Origin of the quiescent centre in *Zea mays*. New Phytol 80: 409–419.

22. Denhardt DT, Edwards DR and Parfett CLS (1986) Gene expression during the mammalian cell cycle. Biochim Biophys Acta 865: 83–125.

23. Domec C, Garbay B, Fournier M and Bonnet J. (1990) cDNA library construction from small amounts of unfractionated RNA: association of cDNA synthesis with polymerase chain reaction amplification. Anal Biochem 188: 422–426.

24. Feldman LJ and Torrey JG (1975) The quiescent center and primary vascular tissue pattern formation in cultured roots of *Zea*. Can J Bot 53: 2796–2803.

25. Feldman LJ (1976) The *de novo* origin of the quiescent centre in regenerating root apices of *Zea mays*. Planta 128: 207–212.

26. Feldman LJ and Torrey JG (1976) The isolation and culture *in vitro* of the quiescent center of *Zea mays*. Am J Bot 63: 345–355.

27. Feldman LJ (1984) Regulation of root development. Annu Rev Plant Physiol 35: 223–242.

28. Fisher DB (1968) Localization of endogeneous RNA polymerase activity in frozen sections of plant tissues. J Cell Biol 39: 745–749.

29. Goyal V and Pillai A (1986) Anatomical and histochemical studies of root apical meristems of some angiosperms. Beitr Biol Pflanzen 62: 167–178.

30. Gurr SJ and McPherson MJ (1991) PCR-directed cDNA libraries. In: PCR: A Practical Approach. MJ McPherson, P Quirke and GR Taylor (eds) pp. 147–170. Oxford University Press, Oxford.

31. Jensen WA (1958) The nucleic acid and protein content of root tip cells of *Vicia faba* and *Allium cepa*. Exp Cell Res 14: 575–583.

32. Jepson I, Bray J, Jenkins G, Schuch W and Edwards K (1991) A rapid procedure for the construction of PCR cDNA libraries from small amounts of plant tissue. Plant Mol Biol Report 9: 131–138.

33. Kodama H, Masaki I, Hattori T and Komamine A (1991) Isolation of genes that are

preferentially expressed at the G1/S boundary during the cell cycle in synchronised cultures of *Catharanthus roseus* cells. Plant Physiol 95: 406–411.

34. McKinney JD and Heintz N (1991) Transcriptional regulation in the eukaryotic cell cycle. TIBS 16: 430–435.
35. Osley MA (1991) The regulation of histone synthesis in the cell cycle. Annu Rev Biochem 60: 827–861.
36. Raghavan V and Olmedilla A (1989) Spatial patterns of histone mRNA expression during grain development and germination in rice. Cell Diff Dev 27: 183–196.
37. Raghavan V (1990) Origin of the quiescent center in the root of *Capsella bursa-pastoris* (L.) Medik Planta 181: 62–70.
38. Romer AS and Parsons TS (1986) The vertebrate body plan. 6th edn. Saunders, Philadelphia.
39. Sabelli PA and Shewry PR (1991) Characterization and organization of gene families at the *Gli-1* loci of bread and durum wheats by restriction fragment analysis. Theor Appl Genet 83: 209–216.
40. Saiki RK, Gelfand DH, Stoffel S, Scharf SJ, Higuchi R, Horn GT Mullis KB and Erlich HA (1988) Primer-directed enzymatic amplification of DNA with a thermostable DNA polymerase. Science 239: 487–491.
41. Sargent T and Dawid I (1983) Differential gene expression in the gastrula of *Xenopus laevis*. Science 222: 135–139.
42. Steeves TA and Sussex IM (1989) Patterns in plant development. 2nd edn. Cambridge University Press, Cambridge.
43. Torrey JG (1972) On the initiation of organization in the root apex. In: The Dynamics of Meristem Cell Populations. MW Miller and CC Kuehnert (eds) pp. 133–147. Plenum Publishing Corporation, New York.
44. Webster PL and Langenauer HD (1973) Experimental control of the activity of the quiescent centre in excised root tips of *Zea mays*. Planta 112: 91–100.

9. Key components of cell cycle control during auxin-induced cell division

DÉNES DUDITS, LÁSZLÓ BÖGRE, LÁSZLÓ BAKÓ, DAMLA DEDEOGLU, ZOLTÁN MAGYAR, TAMÁS KAPROS, FERENC FELFÖLDI and JÁNOS GYÖRGYEY

Abstract

This review article provides a comprehensive summary of the basic molecular and cellular events underlying the induction of the cell division cycle in auxin-treated somatic plant cells. Various pathways of signal transduction chains are discussed as mediators between auxin receptors and alteration of the gene expression pattern. The central role of calcium as a second messenger is analyzed in relation to its interaction with calmodulin and a variety of protein kinases. Experimental data indicate that the control of the cell cycle in higher plants involves several key elements and regulatory mechanisms common to other eukaryotic cells. Recent results show a complex formation between p34^{cdc2} kinase and cyclin-like proteins. Furthermore, the cell cycle-dependent changes in the p34^{cdc2} kinase activity which peak at S- and G2/M-phases suggest functional roles for S- and M-forms of the p34^{cdc2} or related kinases. The homologues of cdc2 and cyclin genes have been cloned from different plant species. The expression of plant cdc2 genes is under transcriptional control in auxin-reactivated cells while high constitutive expression of this gene was found in fast cycling cells grown in suspension culture.

1. Introduction

Auxins are the most intensively studied of the growth hormones in plants. They influence a variety of cellular processes including cell elongation, cell division, differentiation and morphogenesis (for review see [20, 30, 116]. The key role of auxins, especially the synthetic auxin, 2,4-dichlorophenoxyacetic acid (2,4-D), has been demonstrated in the initiation and maintenance of cell division by *in vitro* tissue culture experiments. Exogenous 2,4-D treatment can trigger cell division in somatic cells of inoculated primary explants from various plant organs or in single cells derived from protoplasts. The presence of auxins in the culture medium, with or without cytokinins can initiate either formation of callus tissues or development of organs, or alternatively, somatic embryos from the *in vitro* cultured cells (for review see [19, 36]). Induction of the cell division cycle in hormone-treated cells is a consequence of the activation of a multicomponent cascade system that includes binding of the hormone to receptor molecules, triggering of the signalling pathway, reprogramming of

111

J.C. Ormrod and D. Francis (eds.), Molecular and Cell Biology of the Plant Cell Cycle, 111–131.
© 1993 *Kluwer Academic Publishers. Printed in the Netherlands.*

gene expression and structural reorganization of the cyto-architecture. In this review, we focus mainly on some selected molecular components that are involved in altering the proliferative state of plant cells after application of exogenous hormones.

2. Cell division as an auxin response mediated by activation of a signal transduction chain

A large body of experimental results has been accumulated that indicate the requirement of phytohormones for division of cells cultured *in vitro*. The stimulation of DNA synthesis and the induction of vigorous cell division by auxins can lead to the development of callus tissues from a variety of explants [98, 158, 159]. For example, in leaf tissues of cereals the ability to respond to 2,4-D depends on the developmental stage of inoculated leaf segments. The basal, meristematic segments of wheat leaves can form callus at a low concentration of 2,4-D, whereas high concentrations of auxin are required for G1- to S-phase transition in more distal segments. In the unresponsive zones, nuclear DNA replication can occur but the mitotic cycle is incomplete ([155]; also see PCL John *et al.*, this volume). In maize, the capacity for division and forming callus was restricted to a 4 cm zone at the base of young leaves [154].

Maize protoplasts isolated from the meristematic region of leaves were able to initiate DNA synthesis in culture medium supplemented with auxins [152]. In tobacco protoplast cultures, both auxin and cytokinin were required for the entry into the S-phase and mitosis [27]. Van 't Hof and Kovács [149] proposed two Principal Control Points at the start of the S-phase and mitosis in the regulation of the cell cycle in cultured root tissues.

While a wide range of tissue culture studies supports the central role of auxins in reactivation of the division cycle in cells of primary explants, the direct involvement of auxins in the specific control of the cell cycle has not been fully established in long-term suspension cultures. The experimental data are highly variable depending upon the type of culture used (for a review see [52]). Auxin-depletion arrested the cells at G1-phase in carrot cultures [108]. In contrast, 2,4-D starvation did not arrest sycamore cells at a specific point in the cell cycle [41]. In cultured carrot cells, grown in the presence of 2,4-D, the duration of G1-phase was found to be considerably longer than *in vivo* [10]. The potential for reactivation of the cell cycle in cultured plant cells reflects an open, flexible developmental programme in higher plants *in vivo* that is based on the meristematic nature of various tissue regions each exhibiting a capacity for continuous cell division. In this respect, considerable differences can be detected between cells of dicot and monocot plants. The concept of plastic growth and development (see [147, 138]) might be important in relation to a general principle of control of cell division and differentiation in higher plants. Specificity and the unique nature of plant cell differentiation is such that somatic cells can become totipotent during hormone-induced cell divisions and, through formation of somatic embryos, the whole developmental programme is restarted in cultures *in vitro* (reviewed in [36]). The potential for re-entry into the cell cycle and the transition from a somatic to an embryogenic cell-type is a genetically determined trait that is related to auxin sensitivity of the cells.

We will now focus on the mode of auxin action with respect to the primary molecular and biochemical changes, which alter gene expression or lead to a stimulation of cell proliferation. Several recent analyses have considered auxins as external signals received by a receptor system and subsequently mediated by second messengers in a transduction chain (for review see [2, 16, 54, 103, 116, 121, 135]). The availability of this new information about the various components of the perception and internal transmission of auxin signals can also help to identify important molecular events that lead to the division of cells

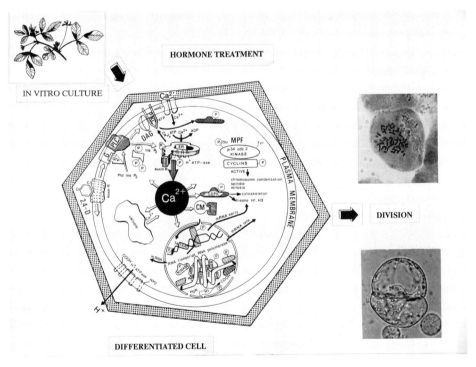

Fig. 1. Scheme postulating the components in an auxin (2,4-D) activated signal transduction cascade involved in the reprogramming of gene expression and induction of cell division.
R: receptor; G: GTP-binding protein; PLC: phospholipase C (phosphoinositidase); Ptd InsP$_2$: phosphatidylinositol 4,5-bisphosphate; InsP$_3$: inositol 1,4,5-trisphosphate; ER: endoplasmic reticulum; DAG: 1,2-diacylglycerol; PKC: protein kinase C; P: phosphoprotein; CDPK: Ca^{2+} dependent protein kinase; CM: calmodulin; MPF: maturation promoting factor; histones: H1; H2A; H2B; H3; H4. References are cited in text. Some components of this presentation are from Sussman and Harper [142]; Morse *et al.* [103] Csordás [28].

exposed to hormones *in vitro*. Figure 1 presents a simplified scheme concerning some key elements of the complex network of the biochemical changes involved in auxin activated cell division. It is now well established that auxins interact with receptor molecules or auxin-binding proteins located in the plasmalemma, endoplasmic reticulum and tonoplasts (reviewed by [73, 106, 116, 117]). Using anti-idiotypic antibodies, Prasad and Jones [122] have identified two putative auxin receptors as 65 and 67 kDa auxin-binding proteins in the nucleus. The nuclear localization of these proteins might have functional significance for the transcriptional control of auxin-regulated genes, although there are few hormone-binding proteins data in plants. Barbier-Brygo *et al.* [8] showed a correlation between auxin sensitivity and the level of tobacco proteins immunologically related to auxin-binding proteins from maize coleoptiles. Conformational changes caused by the binding in auxin-binding proteins of maize may have functional consequences [106]. Plasma membrane H^+-ATPase can serve as an effector-protein by mediating the auxin-induced hyperpolarisation of the membrane. Antibodies against auxin-binding protein from maize can block this response [7]. In addition, the altered H^+-ATPase activity is responsible for H^+ extrusion and change of cytoplasmic pH [132] and auxins can induce pH oscillations [45]. The activation of the cell division cycle requires an increase of cytoplasmic pH in fertilized animal eggs [35]; oscillations of intercellular pH could be a component of cell cycle control (for a review see [53]).

The interaction of hormone with binding protein can stimulate further steps in the signalling pathway including the guanine nucleotide binding protein (G-protein)-dependent activation of phospholipases. The first evidence for the presence of G-proteins in plants was obtained by biochemical analysis. GTP-binding and hydrolysis were detected in membrane fractions from various plant species (for reviews see [54, 116, 135]). Genes encoding members of the *ras*-related *ypt* subfamily were also isolated (see review [116]). Furthermore, cDNA clones of a G protein α subunit from *Arabidopsis* and tomato have been identified [91, 92]. Phospholipases (A_2; C; D) as major phospholipid-degrading enzymes have been detected in plants and their function in the phosphatidylinositol cycle (IP cycle) has been widely analyzed [54, 83, 103]. So far, there is only circumstantial evidence for changes in turnover of inositol phosphates and phosphatidylinositols as a consequence of auxin treatment or transition from the non-proliferative state to cell division [39, 59, 60, 102, 160]. Coupling of auxin (e.g. 2,4-D) induced activation of phospholipase A_2 with auxin receptor function and the requirement for the presence of GTP was shown by Scherer *et al.* [135] and André and Scherer [3]. The link between the metabolism of phosphatidylinositol 4,5-bisphosphate and the calcium signaling is established through the increase of cytosolic inositol triphosphate and associated mobilization of intracellular calcium ([34, 127, 130, 136], see also Fig. 1, also see DE Hanke this volume).

In elucidating the responses of mammalian cells to growth factors it has become evident that there is another branch of second messenger system related to the IP cycle [11]. The product of phosphatidylinositol 4,5-bisphosphate hydrolysis is 1,2-diacylglycerol (DAG) which remains in membranes and initiates the activation of protein kinase C (PKC). Protein kinase C is a serine/threonine-specific protein kinase that requires Ca^{2+} and phospholipid, particularly phosphatidyl serine, for its activation (for reviews see [67, 109]). DAG significantly increases the affinity of PKC for Ca^{2+} but has a much reduced affinity for phosphatidyl serine. Tumour-promoting phorbol esters with structural similarity to DAG also activate PKC both *in vitro* and *in vivo* [21, 72]. Analysis of the complete amino acid sequence of PKC revealed the structural and functional domains of this kinase. PKC is represented by different isozymes ([113]; see the reviews by [67, 109]). Activation of the PKC-dependent signal pathway can be responsible for long-term cellular functions required for the change of gene expression and the initiation of cell proliferation. Transcription factor AP1 consisting of heterodimer of fos and jun oncoproteins is modified upon PKC activation [6]. Detection of the regulatory role of PKC in many different cellular responses including mitogenetic activation in animal cells has focused attention on resolving an analogous signaling pathway in plants. At present, critical demonstration of plant protein kinase with identical structural and functional characteristics to the animal PKC has not yet been reported. However, it seems likely that a similar regulatory system with analogous elements may exist in plants. Both biochemical data and sequence analysis of various plant protein kinase genes suggest a role for a signaling pathway like the PKC mediated one. Clearly, several Ca^{2+}/phospholipid-stimulated plant protein kinases have been found and partially characterized [38, 43, 90, 95, 133]. Moreover, there is a report about the phorbol-ester activation of a wheat kinase [114], but DAG activated kinase has not yet been found in plants. One of the Ca^{2+}-dependent plant protein kinases was able to phosphorylate a synthetic oligopeptide known to be a specific substrate for the animal PKC [115].

Molecular cloning and structural analysis of various plant kinase genes can also provide information about similarities and differences between the key components of signalling based on phosphorylation in plant and animal cells. Lawton *et al.* [80] have identified bean and rice cDNAs encoding for serine/threonine kinases with catalytic domains that are closely related to cyclic nucleotide-dependent protein kinases and the PKC family. The regulatory domain of these plant kinases did not show homology to other known kinases. Lin *et al.* [88] described a protein kinase cDNA clone from pea that showed 69.6-76.1% similarity and 45-49% identity to PKC-β. Further attempts in cloning of plant genes, homologous to the PKC gene family and characterization of protein kinases, with similar functional characteristics and substrate specificity are needed for testing the potential role of a signalling pathway based on a PKC-like plant kinase. Clearly, it will be important to determine the possible involvement of this group of protein kinases in relation to auxin action which is required for cell division.

3. Calcium centred messenger system and its interaction with a protein kinase network

An important concept that has emerged from studies on the molecular and physiological consequences of auxin action is the central role of calcium as a second messenger [20, 46, 54, 103, 121, 146]. The rise in cytosolic Ca^{2+} level can be a transiently acting signal that is generated either through the IP cycle (see before) or modification of Ca^{2+} channel activity [137]. Various effects evoked by calcium fluxes are mediated by Ca^{2+} binding proteins e.g. calmodulin or a variety of protein kinases. Calmodulin (CM), is a ubiquitous intracellular Ca^{2+} receptor that can form a complex with various enzymes or proteins after conformational changes caused by binding of calcium [94, 118, 131]. Plant calmodulins are small, acidic proteins of molecular weight 17 kDa [145]. Calmodulin has been detected in plasma membrane preparation [26] as well as in nuclei in the spindle pole region close to the kinetochore fibres [79]. Calmodulins possess four Ca^{2+}-binding domains of approximately 148 amino acid residues with high sequence conservation that have been identified in deduced amino acid sequence of cDNAs encoding for barley and potato calmodulin [68, 89]. High levels of calmodulin mRNA were found in strawberry fruits and in stolon tips of potato; furthermore, auxin treatment has resulted in increased transcript levels [68]. Experiments with mammalian cells suggested the involvement of calmodulin in the control of progression through the cell cycle at G1/S boundary [22, 23]. During the entry into S-phase there is a rapid increase in calmodulin that is preceded by an increase in calmodulin mRNA [23]. In the pea root apex the highest concentration of calmodulin was found in areas of cell division and the lowest in the region of cell expansion [1]. Also in embryogenic axes of *Cicer arientinum* L. zones of active cell division showed elevated CM levels [62]. Association of calmodulin with the mitotic spindle and with the phragmoplast was detected by double immunofluorescence labelling of onion cells at cytokinesis [156]. Ca^{2+} is considered to be a regulator of mitosis and cytokinesis (see review [61]).

One of the principal mechanisms by which Ca^{2+} acts to regulate cellular functions is the activation of Ca^{2+} or calmodulin dependent protein kinases. Phosphorylation of the amino acid residues, serine, threonine and tyrosine is a reversible, covalent modification. Transfer of a phosphate moiety from a nucleoside triphosphate regulates the biological activity of proteins, and regulation of metabolism in plant cells by protein kinases and phosphatases is becoming increasingly apparent (for reviews see [116, 121, 126, 146]. The early experiments studying phosphorylation indicated that auxins can enhance the phosphorylation of nuclear proteins [105, 134]. Moreover, a relationship between cell division and phosphorylation of nuclear proteins was indicated in several experiments with various cell types. Arfmann and Willmitzer [4] detected a five-fold increase in endogenous protein kinase activity in nuclei from actively dividing cells than in nuclei of quiescent cells in leaves of tobacco. In cell suspension culture, protein kinase activity increased after the lag phase when cells showed a high rate of protein synthesis and rapid proliferation [14]. The chromatin-bound protein kinases were activated during *Agrobacterium*-induced tumour development on potato tuber discs and the pattern of chromatin phosphoproteins was highly reproducible and stage-specific [71]. However, in relation to auxin-induced changes in protein phosphorylation the experimental data did not support a generally acceptable conclusion. Treatment of coleoptiles of a monocotyledonous plant, oat, with naphthaleneacetic acid (NAA) did not cause alteration in phosphorylation *in vivo*. Ca^{2+} and calmodulin-dependent phosphorylation was observed in both control and auxin-treated tissues [150]. In contrast, in pea epicotyls phosphorylation of selected proteins was either increased or decreased after treatment with indoleacetic acid. The authors suggest the possible involvement of calmodulin in auxin-induced phosphorylation changes [128]. Other phytohormones such as cytokinins or ethylene can also modify the phosphorylation of proteins e.g. in cabbage leaf discs or artichoke explants [75, 125]. In addition, stresses including heat shock or digestion of cell walls by driselase trigger rapid changes in protein kinase activity [13, 77].

In general, the Ca^{2+}-dependent protein kinases require effector molecules (either calmodulin or phospholipids) in addition to Ca^{2+} for activation. A Ca^{2+}- and calmodulin activated autophosphorylating kinase of molecular weight 18kDa has been characterized in the plasma membranes of pea [12]. Ca^{2+}- or Ca^{2+}/calmodulin-dependent kinase activity in membrane fractions can modulate essential functions e.g. H^+ ATPase activity [78, 161]. Above, we listed publications that provided evidence for Ca^{2+}/phospholipid stimulated protein kinases in plants. In addition to these two major classes of protein kinases, Ca^{2+}-dependent and calmodulin-independent kinase activity was also detected in different plant species [9, 15, 56, 74, 125]. The soybean Ca^{2+}-dependent kinase (CDPK) with a molecular weight of 52 kDa was not activated by phospholipids. *In vitro* this kinase phosphorylates histone III-S, myosin light chain, ribosomal proteins and has autophosphorylation capability [124]. Based on structural features predicted from the deduced amino acid sequence of the cloned cDNA, this CDPK shares homology with calmodulin-dependent kinases in the catalytic domain and possesses a calmodulin-like region with four calcium-binding domains [57]. A homologue of the soybean CDPK cDNA has been cloned from carrot [140]. The biological function of soybean CDPK can be related to the cytoskeleton since this kinase was co-localized with F-actin in plant cells by using monoclonal antibodies ([57]; also see PCL John *et al.* and Francis and Herbert, this volume).

A Ca^{2+}-dependent protein kinase with molecular weight of 52-54 kDa was partially purified from cultured alfalfa cells. Exhibiting similarities to the soybean CDPK, the alfalfa kinase was capable of autophosphorylation but calmodulin failed to activate the enzyme preparation after the purification steps [15, 115]. However, the functional role of this alfalfa CDPK is not fully understood, although the considerable increase in the autophosphorylation activity in protoplast cultures at the time of the first cell division suggests a

118

relation between this CDPK and cell cycle reactivation. As shown by Fig. 2A, three major phosphoproteins can be recognized after *in vitro* phosphorylation using cell extracts from alfalfa cells at 4 days after protoplast isolation. The 66 kDa phosphoprotein could be detected in the presence or absence of calcium. The 52-52 kDa proteins, corresponding to the autophosphorylated forms of alfalfa CDPK, showed Ca^{2+}-dependence with a maximum signal occurring in cell extracts from cultures that already possessed protoplast-derived dividing cells. In this experiment, only the major phosphoproteins could be recognized because phosphatase inhibitors were not added to the reaction mixture. In addition, cultures grown on a 2,4-D-supplemented medium comprised dividing cells, which exhibited a distinct substrate-specificity in the *in vitro* histone phosphorylation assays (Fig. 2B). Separation of phosphorylated histones by acid urea-gel analysis TritonX-100 gradient (described in [153]) revealed the preferential phosphorylation of histone H3. At present, we do not know whether histone H3 is an *in vivo* substrate for this CDPK. Phosphorylation of histone H3 is one of the first responses during mitotic activation of animal cells. Further functional characterization of various plant CDPK kinases can be achieved through the production of antibodies and by the expression of cloned cDNAs in transgenic plants.

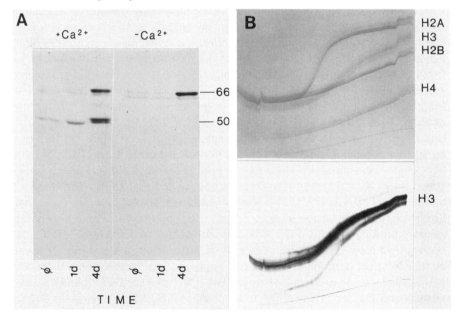

Fig. 2. Increased autophosphorylation activity of the alfalfa Ca^{2+}-dependent protein kinase (CDPK) in cell extracts from dividing cells in protoplast culture.

A: *In vitro* phosphorylation of proteins in cell extracts from cultures at the 1st and 4th days after isolation of protoplasts. The phosphorylation conditions are described by Bögre *et al.* [15].

B: Histone phosphorylation by partially purified alfalfa CDPK. Separation of various histones was carried out according to [153]. This experiment was carried out with the help of JH Waterborg.

Both the early and late responses to the auxin treatment involve significant transcriptional changes (see T Nagata and Y Takashi, this volume). A large variety of auxin-induced genes have been cloned and characterized [50, 55, 96, 97, 144]. An auxin-regulated tobacco cDNA with expression at G0/S transition was cloned by Takahashi *et al.* [143]. Considering several examples, it can be hypothesized that phosphorylation of transcriptional factors by specific kinases at the end of the signalling network may play a major role in regulating expression of genes involved in auxin induced cell division. Induction of fos protooncogene is controlled by phosphorylation of the Serum-Responsive-DNA binding Factor (*SRF*) [123]. The factor CREB that binds the cAMP-responsive element is also under the regulatory system based on phosphorylation [157]. In plants, the DNA-binding ability of AT-1 protein was shown to be reversibly modulated by phosphorylation [29]. In the future, a better understanding of the link between the hormone activated protein kinase network and the regulation of transcription should reveal the molecular basis of auxin triggered cellular changes including cell division.

4. General model of cell cycle control: role for p34^{cdc2} kinase and cyclins in regulation of plant cell division in relation to auxin action

Over the past few years, discoveries of highly conserved processes and controlling elements in cell division from yeast to man have supported a unifying hypothesis for regulation of the cell cycle in eukaryotes (see J Hayles and P Nurse, this volume). A central feature of the current models is a 34 kDa serine/threonine kinase encoded by the *cdc*2 gene in *S. pombe* or by the CDC28 gene in *S. cerevisiae* and different homologous genes in other eukaryotic cells (reviewed by [24, 32, 110, 11]). The cell cycle-dependent function of p34^{cdc2} kinase is under the control of phosphorylation/dephosphorylation at defined tyrosine or serine/threonine residues of the kinase and is also regulated by complex formation with a distinct type of cyclin [24, 31, 33, 49, 76, 139]. The p34^{cdc2} kinase is required at two points in the cell cycle, the transition from G2 to M and the transition from G1 to S [18, 82, 112, 120, 129, 141]. In *in vitro* phosphorylation experiments many proteins are substrates for the p34^{cdc2} kinase. Proteins responsible for the major structural changes in mitosis can also be phosphorylated by this kinase (for reviews see [32, 87, 101]). For example, *cdc*2 kinase regulates the changes in microtubule dynamics occurring at the interphase-metaphase transition in cell-free extracts from *Xenopus* eggs ([151; also see D Francis and RJ Herbert, this volume).

In addition to post-translational modifications, the activity of p34^{cdc2} is regulated by its interaction with cyclins, proteins of molecular size 45-60 kDa, first identified in marine invertebrates and later in a wide variety of species including man [40, 100, 119, 129, 139]. Various members of the family of cyclins or cyclin-like proteins are positive effectors of cell cycle progression by activating *cdc*2 or related protein kinases. The G2/M-phase transition is controlled by an activity termed M-phase or maturation promoting factor (MPF) with kinase function. MPF consists of two subunits: the catalytic subunit (p34[cdc2] kinase) and a regulatory subunit (cyclin B) (for a review see Dorée [31]). Galaktionov and Beach [48] suggested a multifunctional role for B-type cyclins: regulating substrate specificity by complexing with cdc2 kinase see J Hayles and P Nurse, this volume activated by the cdc25 tyrosine phosphatase.

G1-specific cyclins (for review see [129]) whose function is required for the G1/S-transition were first identified in S. cerevisiae as cyclin-like proteins (Cln1, Cln2, Cln3). A Drosophila and a human gene encoding a protein with G1 cyclin activity in cross-species complementation of CLN deficiency in budding yeast have also been identified [85, 86]. Cyclin A shares common sequence motifs and functional similarity with cyclin B but it has a specific regulatory role that can be linked to multiple check points in the cell cycle. In Xenopus, the complex between cyclin A and p34[cdc2] showed an earlier increase in histone H1 kinase activity in comparison with cyclin B-p34[cdc2] complexes [99]. In Drosophila, cyclin A plays a unique and essential role in the G2 to M transition [84]. Association of p60, cyclin A with a p34[cdc2]-related protein [33 kDa] was detected in S-phase whereas p62 cyclin B was complexed with p34[cdc2] later in G2-phase in HeLa cells [120]. In human cells, cyclin A can associate with p34[cdc2] as well as with a 33 kDa cdc2-related protein (cyclin-dependent protein kinase) encoded by cdk2 gene [37, 107, 148]. The accumulation of the cyclin A/cdk2 complex showed that the protein kinase activity accompanies and follows S-phase [66, 104]. Cyclin A has been found to interact with E2F transcriptional factor and the complex also contains p33[cdk2] kinase and Rb-related p107 protein during S-phase [42].

From the above, it is clear that significant progress has been achieved in the understanding of some key regulatory steps in cell cycle control of eukaryotic cells. Similar studies on plant cells are now in agreement with the concept of a universal regulatory system in cell cycle control in eukaryotes. In certain cases analogous controlling elements have already been proven to exist in higher plants. Several laboratories have produced Western blot data showing the presence of p34[cdc2]-like protein in cell extracts from different plant species. Antibodies raised against either the highly conserved internal PSTAIR peptide or p34 from the fission yeast (Mab-J4) cross-reacted with protein of 34 kDa molecular weight [5, 44, 65, 69]. In some plants, a protein-doublet was recognized by the antibodies at a molecular weight of approximately 34 kDa. The level of p34[cdc2]-like protein varied according to the frequency of dividing cells in tissues such as wheat leaf or carrot cotyledon [51, 70]. In carrot explants, 2,4-D treatment increased the amounts of p34[cdc2]-like protein, possibly through activation of cells as they entered the cell cycle. After Sephacryl S-200 gel filtration chromatography of the pea extract, cross-reactive p34 was detected in both low and high molecular mass fractions. Feiler and Jacobs [44] suggested that p34 is present as a monomer and as a component of a complex. The histone H1 kinase activity was largely confined to the high molecular mass fractions. The level of p34^{cdc2}-specific histone H1 kinase activity was determined by p13-Sepharose affinity binding in protein extracts from maize tissues [25]. The *suc*1$^+$ gene encodes a protein of 13 kDa that was shown to be associated in a complex with the *cdc*2 protein [17]. The p34^{cdc2} bound specifically to p13^{suc1}-Sepharose and was active as a protein kinase. By precipitation with p13^{suc1}-Sepharose beads, high H1 kinase activity was detected in maize apical meristems, while extracts from maize leaves did not show such protein kinase function [25].

The first indication of the presence of a cyclin-like protein in plants was provided by Bakó *et al.* [5]. During DEAE-Sepharose chromatography of proteins from alfalfa suspension culture, a 62 kDa protein was recognized by antibodies raised against cyclin B encoded by the *cdc*13 gene from fission yeast. Association of an alfalfa homologue of p34^{cdc2} with a cyclin-like protein was demonstrated by two experimental approaches [93]. Western blots of proteins bound to p13^{suc1}-Sepharose indicated the presence of a 34 kDa doublet recognized by anti-PSTAIR antibodies and two other proteins cross-reacted with human cyclin A antibodies. The mobility of the proteins detected by cyclin A antibodies in SDS-PAGE corresponded to proteins with molecular mass of 65 and 32 kDa. Furthermore, the cyclin A antibodies immunoprecipitated a histone H1 kinase activity from synchronized alfalfa cells with an activity peak at early S-phase. Analysis of cultured alfalfa cells synchronized by hydroxyurea treatment revealed the highest histone H1 kinase activity in the p13^{suc1}-Sepharose-bound protein fractions from cells at S-phase. A second peak with lower kinase activity was observed at G2/M-phase. Detection of periodic changes in histone phosphorylation by p34^{cdc2} kinase with maximum activity at S- and G2/M-phase suggests functional roles for S- and M-forms of p34^{cdc2} in plants. The coincidence of elevated p34^{cdc2} histone H1 kinase activity with increase in incorporation of [^{3}H]-thymidine in cultured leaf mesophyll protoplasts also suggests the involvement of p34^{cdc2} kinase in reactivation of the cell cycle (Fig. 3D).

The extensive degree of homology in amino acid sequences at highly conserved functional domains of both p34^{cdc2} and cyclins (for reviews see [32, 81, 129]) has enabled cloning strategies for the isolation of plant homologues of these genes [25, 44, 47, 58, 63, 65]. Comparative sequence analysis of different *cdc*2 homologues from plants revealed a 60-70% homology with other eukaryotic *cdc*2/*CDC*28 genes. Moreover, plant *cdc*2 cDNAs could functionally complement *cdc*2 mutations in fission yeast [25, 47, 65]. The sequence conservation is higher within the plant kingdom but *cdc*2 variants can also be recognized amongst the *cdc*2 homologues. Regarding the expression of plant homologues to the *cdc*2 gene, differences in the level of *cdc*2 transcripts were detected in various tissues of maize, *Arabidopsis* and alfalfa [25, 47, 65]. A more comprehensive analysis of expression of alfalfa *cdc*2 gene during the cell cycle and during the activation of cell division suggested a proliferative state-dependent regulation in *cdc*2 transcript level (Fig. 3B). Both in 2,4-D-treated root explants, and in cultured mesophyll protoplasts, transcriptional

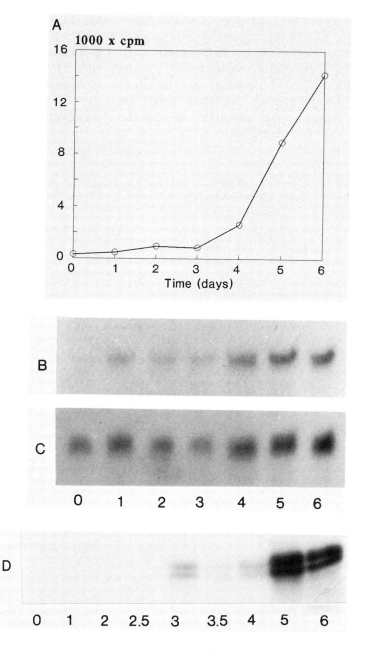

Fig. 3. Activation of alfalfa *cdc2* gene and increased p34[cdc2] kinase activity during reactivation of the mesophyll protoplast-derived cells in S-phase of the cell cycle [93].
A: Incorporation of [³H]-thymidine into cells from mesophyll protoplasts
B: Northern hybridization with the alfalfa *cdc2* gene
C: Northern hybridization with Msc27 cDNA probe as a reference
D: Histone H1 phosphorylation by protein fractions-bound to p34[cdc2] Sepharose beads.

activation was detected in parallel with increased DNA synthetic activity indicating the re-entry of cells into the cycle ([93], see also Fig. 3B). In contrast, in rapidly cycling alfalfa cells maintained in suspension culture, the accumulation of *cdc*2 transcripts was independent of the cell cycle phase at which the cells were analyzed.

As with the *cdc*2 genes, molecular cloning also provided evidence for the existence of cyclin-like proteins in plants. Hata *et al.*[58] have identified carrot and soybean cDNAs which proved to be homologues of known cyclin genes. The amino acid sequences reflected conserved regions with a similarity to both A-type and B-type cyclins. However, unusual amino acid sequence elements were found at the central portion of the cyclin box. The level of cyclin mRNAs was higher in somatic embryos from carrot and in tissues with dividing cells in soybean seedlings. Different cDNAs with homology to cyclin B genes were also cloned and characterized in alfalfa [64].

5. Outlook

It is a challenging task to combine the recent results of biochemical and physiological studies with information gained from molecular cloning work, in order to get new insights into the old problem: how do auxins regulate plant growth and development? Especially, at present, there is a limitation in our understanding of the molecular background of the long-term auxin actions, such as activation of cell division and support of the continuous proliferation of cells either in *in vitro* culture or in meristematic tissues. Several lines of evidence support the functional role of similar components in signal transduction and cell cycle control in eukaryotic cells, including plants. Recognition of the conserved cellular functions, such as cell division, can help to outline research strategies for studies on cell cycle control in higher plants. As shown by the first few examples, we can expect considerable progress in identification of the conserved elements either in the biochemical pathways or in the structure and function of genes. The cross-species complementation, using yeast mutants, can help significantly in the identification of essential genes. At the same time, it is predictable that the understanding of plant specific features will require a more comprehensive research effort both in the analytical and in the cloning work. As far as the various experimental systems are concerned, it may be important to emphasize the differences between cells as functional units, in tissues *in vivo* and cells cultured *in vitro*. In the latter case, the hormonal effects are closely linked with stress-induced changes. At the present time, the identification of the key components in the regulation of the division cycle attracts most attention. The understanding of the functional coordination between various elements and processes may be advanced later by the use of transgenic plants. Transformants, with cell cycle genes altered in their regulatory or coding regions, will provide a very powerful experimental tool for such functional analyses. This approach should prove to be important in revealing the relationship between the primary molecular events and the dramatic structural re-organization of cells as a consequence of cell division. The advanced methodology of biochemistry, molecular and cell biology brings new horizons to studies of plant cell division. This new information is essential in order to outline a general concept about the control of the eukaryotic cell cycle and may open the way to modify plant growth and development.

Acknowledgement

The authors thank Ildikó Zsigó for excellent work in preparation of the manuscript.

References

1. Allan E and Trewavas A (1985) Quantitative changes in calmodulin and NAD kinase during early cell development in the root apex of *Pisum sativum* L. Planta 165: 493-501.
2. Anderson JM (1989) Membrane-derived fatty acids as precursors to second messengers. In: Second Messengers in Plant Growth and Development. WF Boss and Morré DJ (eds), pp. 167-179 New York: Alan R. Liss.
3. André B and Scherer GFE (1991) Stimulation by auxin of phospholipase A in membrane vesicles from an auxin-sensitive tissue is mediated by an auxin receptor. Planta 185: 209-214.
4. Arfmann HA and Willmitzer L (1982) Endogenous protein kinase activity of tobacco nuclei. Comparison of transformed, non-transformed cell cultures and the intact plant of *Nicotiana tabacum*. Plant Sci Lett 26: 31-38.
5. Bakó L, Bögre L and Dudits D (1991) Protein phosphorylation in partially synchronized cell suspension culture of alfalfa. In:NATO Adv Studies on Cellular Regulation by Protein Phosphorylation. L Heilmayer (ed). H56: 435-439 Berlin: Springer Verlag.
6. Barber JR and Verma IN (1987) Modification of fos proteins: phosphorylation of c-fos but not v-fos is stimulated by TPA and serum. Mol Cell Biol 7: 2201-2211.
7. Barbier-Brygoo H, Ephritikhine G, Klambt D and Guern J (1990) The sensitivity of plant protoplasts to auxin is likely modulated by the number of receptors at the plasmalemma. In: Signal Perception and Transduction in Higher Plant. R Ranjeva and A Boudet (eds). NATO ASI Series, Vol. H47 1-12, Berlin: Springer Verlag.
8. Barbier-Brygoo H, Ephritikhine G, Klambt D, Maurel C, Palme K, Schell J and Guern J (1991) Perception of the auxin signal at the plasma membrane of tobacco mesophyll protoplasts. Plant J 1: 83-93.
9. Battey NH and Venis MA (1988) Calcium-dependent protein kinase from apple fruit membranes is calmodulin-independent but has calmodulin-like properties. Planta 176: 91-97.
10. Bayliss MW (1975) The duration of the cell cycle of Daucus carota L. in vivo and *in vitro*. Exp Cell Res 92: 31-38.
11. Berridge MJ (1987) Inositol trisphosphate and diacylglycerol: two interacting second messengers. Ann Rev Biochem 56: 159-193.
12. Blowers DP and Trewavas AJ (1987) Autophosphorylation of plasmamembrane bound calcium and calmodulin-dependent protein kinase from pea seedlings and modification of catalytic activity by autophosphorylation. Biochem Biophys Res Comm 143: 691-696.
13. Blowers DP, Boss WF and Trewavas AJ (1988) Rapid changes in plasma membrane protein phosphorylation during initiation of cell wall digestion. Plant Physiol 86: 505-509.
14. Böcher M, Erdmann H, Heim S and Wylegalla C (1985) Protein kinase activity of *in vivo*

cultured plant cells in relation to growth and starch metabolism. J Plant Physiol 119: 209-218.

15. Bögre L, Oláh Z, Dudits D (1988) Ca^{2+} dependent protein kinase from alfalfa (*Medicago varia*): partial purification and autophosphorylation. Plant Sci 58: 135-144.

16. Boss WF (1989) Second messengers in plant growth and development. In: Plant Biol. WF Boss and DJ Morre (eds). Vol.6. New York: Alan R Liss.

17. Brizuela L, Draetta G and Beach D (1989) Activation of human *cdc*2 protein as a histone H1 kinase is associated with complex formation with the p62 subunit. Proc Natl Acad Sci USA 86: 4362-4366.

18. Broek D, Bartlett R, Crawford K and Nurse P (1991) Involvement of p34[cdc2] in establishing the dependency of S phase on mitosis. Nature 349: 388-393.

19. Brown DCW and Thorpe TA (1986) Plant regeneration by organogenesis. In: Cell Culture and Somatic Cell Genetics of Plants. IK Vasil (ed). Plant Regeneration and Genetic Variability, Vol.3. pp. 49-65, Orlando, New York: Academic Press.

20. Brummell DA and Hall JI (1987) Rapid cellular responses to auxin and the regulation of growth. Plant, Cell and Environment 10: 523-543.

21. Castagna M, Takai Y, Kaibuchi K, Sano K, Kikkawa U and Nishizuka Y (1982) Direct activation of calcium-activated phospholipid-dependent protein kinase by tumor promoting phorbol esters. J Biol Chem 257: 7847-7851.

22. Chafouleas JG, Bolton WE, Hidaka H, Boyd AE and Means AR (1982) Calmodulin and the cell cycle: involvement in regulation of cell cycle progression. Cell 28: 41-50.

23. Chafouleas JG, Lagace L, Bolton WE, Boys AE and Means AR (1984) Changes in calmodulin and its mRNA accompany reentry of quiescent (G0) cells in the cell cycle. Cell 36: 73-81.

24. Clarke PR and Karsenti E (1991) Regulation of p34[cdc2] protein kinase: new insights into protein phosphorylation and the cell cycle. J Cell Sci 100: 409-414.

25. Colasanti J, Tyers M and Sundaresan V (1991) Isolation and characterization of cDNA clones encoding a functional p34[cdc2] homologue from *Zea mays*. Proc Natl Acad Sci USA 88: 3377-3381.

26. Collinge M and Trewavas AJ (1989) The location of calmodulin in the pea plasma membrane. J Biol Chem 264: 8865-8872.

27. Cooke R and Meyer Y (1981). Hormonal control of tobacco protoplast nucleic acid metabolism during *in vitro* culture. Planta 152: 1-7.

28. Csordás A (1990) On biological role of histone acetylation. Biochem J 265: 23-38.

29. Datta N and Cashmore AR (1989) Binding of a pea nuclear protein to promoters of certain photoregulated genes is modulated by phosphorylation. The Plant Cell 1: 1069-1077.

30. Davies E (1987) Action potentials as multifunctional signals in plants: a unifying hypothesis to explain apparently disparate wound responses. Plant, Cell and Environment 10: 623-631.

31. Dorée M (1990) Control of M-phase by maturation-promoting factor. Current Opinion in Cell Biol 20: 269-273.

32. Draetta G (1990) Cell cycle control in eukaryotes: molecular mechanisms of *cdc*2 activation. Trends in Biochem Sci 15: 378-383.

33. Draetta G, Luca F, Westendorf J, Brizuela L, Ruderman J and Beach D (1989) *cdc*2 protein kinase is complexed with both cyclin A and B: evidence for proteolytic inactivation of MPF. Cell 56: 829-838.

34. Drobak BK and Ferguson IB (1985) Release of Ca^{2+} from plant hypocotyl microsomes by inositol-1,4,5-trisphosphate. Biochem and Biophys Res Commun 130: 1241-1246.

35. Dube FT, Schmidt CH, Johnson CH and Epel D (1985) The hierarchy of requirements for an elevated intracellular pH during early development of sea urchin embryos. Cell 40: 657-666.

36. Dudits D, Bögre L, Györgyey J (1991) Molecular and cellular approaches to the analysis of plant embryo development from somatic cells *in vitro*. J Cell Sci 99: 474-484.

37. Elledge SJ and Spottswood MR (1991) A new human p34 protein kinase, CDK2, identified by complementation of a *cdc*28 mutation in *Saccharomyces cerevisiae*, is a homolog of *Xenopus* Eg1. EMBO J 10: 2653-2659.

38. Elliott DC and Skinner JD (1986) Calcium-dependent, phospholipid-activated protein kinase in plants. Phytochem 25: 39-44.
39. Ettlinger C and Lehle L (1988) Auxin induces rapid changes in phosphatidylinositol metabolites. Nature 331: 176-178.
40. Evans T, Rosenthal ET, Youngblom J, Distel D and Hunt T (1983) Cyclin: a protein specified by material mRNA in sea urchin eggs that is destroyed at each cleavage division. Cell 33: 389-396.
41. Everett NP, Wang TL, Gould AR and Street HE (1981) Studies on the control of the cell cycle in cultured plant cells II. Effects of 2,4-dichlorophenoxyacetic acid (2,4-D). Protoplasma 106: 15-22.
42. Faha B, Ewen ME, Tsai LH, Livingston DM and Harlow E (1992) Interaction between human cyclin A and adenovirus E1A-associated p107 protein. Science 255: 87-90.
43. Favre B and Turian G (1987) Identification of a calcium- and phospholipid-dependent protein kinase (protein kinase C) in *Neurospora crassa*. Plant Sci 49: 15-21.
44. Feiler HS and Jacobs TW (1990) Cell division in higher plants: a cdc2 gene its 34-kDa product, and histone H1 kinase activity in pea. Proc Natl Acad Sci USA 87: 5397-5401.
45. Felle H (1988) Auxin causes oscillations of cytosolic free calcium and pH in *Zea mays* coleoptiles. Planta 174: 495-499.
46. Ferguson IB and Drobak BK (1988) Calcium and the regulation of plant growth and senescence. Hort Sci 23: 262-266.
47. Ferreira PCG, Hemerly AS, Villaroel R, Montagu MC and Inze D (1991) The *Arabidopsis* functional homolog of the p34^{cdc2} protein kinase. The Plant Cell 3: 531-540.
48. Galaktionov K and Beach D (1991) Specific activation of cdc25 tyrosine phosphatase by B-type cyclins: evidence for multiple roles of mitotic cyclins. Cell 67: 1181-1194.
49. Gautier J and Maller J (1991) Cyclin B in Xenopus-oocytes: implications for the mechanism of pre-MPF activation. EMBO J 10: 177-182.
50. Gee MA, Hagen G and Guilfoyle TJ (1991) Tissue-specific and organ-specific expression of soybean auxin-responsive transcripts GH3 and SAURs. Plant Cell 3: 419-430.
51. Gorst JR, John CL and Sek FJ (1991) Levels of p34^{cdc2}-like protein in dividing, differentiating and dedifferentiating cells of carrot. Planta 185: 304-310.
52. Gould AR (1983) Control of the cell cycle in cultured plant cells. CRC Critical Rev in Plant Sci 1: 315-344.
53. Grandin N and Charbonneau M (1991) Cycling in intracellular free calcium and intracellular pH in *Xenopus* embryos: possible roles in the control of the cell cycle. J Cell Sci 99: 5-11.
54. Guern J, Ephritikhine G, Imhoff V and Pradier JM (1990) Signal transduction at the membrane level of plant cells. In: Progress in Plant Cellular and Molecular Biology. HJJ Nijkamp, LHW Van der Plas and J Van Aartrijk (eds), pp. 466-479 Dordrecht: Kluwer Academic Publishers.
55. Hagen G. Uhrhammer N and Guilfoyle TJ (1988) Regulation of expression of an auxin-induced soybean sequence by cadmium. J Biol Chem 263: 642-646.
56. Harmon AC, Putnam-Evans C and Cormier MJ (1987) A calcium dependent but calmodulin-independent protein kinase from soyabean. Plant Physiol 83: 830-837.
57. Harper JF, Sussman MR, Schaller GE, Putman-Evans C, Charbonneau H and Harmon AC (1991) A calcium-dependent protein kinase with a regulatory domain similar to calmodulin. Science 252: 951-954.
58. Hata S, Kouchi H, Suzuka I and Ishii T (1991) Isolation and characterization of cDNA clones for plant cyclins. The EMBO J 10: 2681-2688.
59. Heim S and Wagner KG (1987) Enzymatic activities of the phosphatidylinositol cycle during growth of suspension cultured plant cells. Plant Sci 49: 167-173.
60. Heim S and Wagner KG (1989) Inositol phosphates in the growth cycle of suspension cultured plant cells. Plant Sci 63: 159-165.
61. Hepler PK and Wayne RO (1985) Calcium and plant development. Annu Rev Plant Physiol 36: 397-439.
62. Hernandez-Nistal J, Rodrigez D, Nicolas G and Aldasoro JJ (1989) Abscisic acid and

temperature modify the levels of calmodulin in embryogenic axes of *Cicer arietinum*. Plant Physiol 75: 255-260.

63. Hirayama T, Imajuku Y, Anai T, Matsui M and Oka A (1991) Identification of two cell-cycle-controlling *cdc*2 gene homologs in *Arabidopsis thaliana*. Gene 105: 159-165.

64. Hirt H, Mink M, Györgyey J, Pfosser M, Dudits D and Heberle-Bors E (1991) Isolation and characterization of plant cyclin gene. In: 3rd Int Congress of Plant Mol Biol Tucson, No 891.

65. Hirt H, Páy A, Györgyey J, Bakó L, Németh K, Bögre L, Schweyen RJ, Heberle-Bors E and Dudits D (1991) Complementation of a yeast cell cycle mutant by an alfalfa cDNA encoding a protein kinase homologous to p34[cdc2]. Proc Natl Acad Sci USA 88: 1636-1640.

66. Huang S, Lee WH and Lee YHP (1991) A cellular protein that competes with SV40T antigen for binding the retinoblastoma gene product. Nature 350: 160-162.

67. Jaken S (1990) Protein kinase C and tumor promoters. Current Opinion in Cell Biology 2: 192-196.

68. Jena PK, Reddy ASN and Poovaiah BW (1989) Molecular cloning and sequencing over cDNA for plant calmodulin: signal-induced changes in the expression of calmodulin. Proc Natl Acad Sci USA 86: 3644-3648.

68. Jena PK, Reddy ASN and Poovaiah BW (1989) Molecular cloning and sequencing over cDNA for plant calmodulin: signal-induced changes in the expression of calmodulin. Proc Natl Acad Sci USA 86: 3644-3648.

69. John PCL, Sek FJ and Lee MG (1989) A homolog of the cell cycle control protein p34[cdc2] participates in the division cycle of *Chlamydomonas* and a similar protein is detectable in higher plants and remote taxa. The Plant Cell 1: 1185-1193.

70. John PCL, Sek FJ, Carmichael JP and McCurdy DW (1990) p34[cdc2] homologue level, cell division, phytohormone responsiveness and cell differentiation in wheat leaves. J Cell Sci 97: 627-630.

71. Kahl G and Schafer W (1984) Phosphorylation of chromosomal protein changes during the development of crown gall tumors. Plant and Cell Physiol 25: 1187-1196.

72. Kawahara Y, Minakuchi R, Sano K and Nishizuka Y (1980) Phospholipid turnover as a possible transmembrane signal for protein phosphorylation during human platelet activation by thrombin. Biochem Biophys Res Commun 97: 309-317.

73. Klambt D (1990) A view about the function of auxin-binding proteins at plasma membranes. Plant Mol Biol 14: 1045-1050.

74. Klucis E and Polya GM (1988) Localization, solubilization and characterization of plant membrane-associated calcium-dependent protein kinases. Plant Physiol 88: 164-171.

75. Koritsas VM (1988) Effect of ethylene and ethylene precursors on protein phosphorylation and xylogenesis in tuber explants of *Helianthus tuberosus* (L.) J Exp Bot 39: 375-386.

76. Krek W and Nigg EA (1991) Differential phosphorylation of vertebrate p34[cdc2] kinase at the G1/S and G2/M transitions of the cell cycle; identification of major phosphorylation. The EMBO J 10: 305-316.

77. Krishnan HB and Pueppke SG (1987) Heat shock triggers rapid protein phosphorylation in soybean seedings. Biochem Biophys Res Commun 148: 762-767.

78. Ladror US and Zielinski RE (1989) Protein kinase activities in tonoplast and plasmalemma membranes from corn roots. Plant Physiol 89: 151-158.

79. Lambert AM, Vantard M, Van Eldik L and De May J (1983) Immunolocalization of calmodulin in higher plant endosperm during mitosis. J Cell Biol 97: 40a.

80. Lawton MA, Yamamoto RT, Hanks SK and Lamb CJ (1989) Molecular cloning of plant transcripts encoding protein kinase homologs. Proc Natl Acad Sci 86: 3140-3144.

81. Lee MG and Nurse P (1987) Complementation used to clone a homologue of the fission yeast cell cycle control gene *cdc*2. Nature 327: 31-35.

82. Lee MG, Norbury CJ, Spurr NK and Nurse P (1988) Regulated expression and phosphorylation of a possible mammalian cell cycle control protein. Nature 333: 676-679.

83. Lehle L (1990) Phosphatidylinositol metabolism and its role in signal transduction in growing plants. Plant Mol Biol 15: 647-658.

84. Lehner C and O'Farrell PH (1990) The roles of *Drosophila* cyclins A and B in mitotic control. Cell 61: 535-547.
85. Leopold P and O'Farrell PH (1991) An evolutionary conserved cyclin homolog from *Drosophila* rescues yeast deficient in G1 cyclins. Cell 66: 1207-1216.
86. Lew DJ, Dulic V and Reed SI (1991) Isolation of three novel human cyclins by rescue of G1 cyclin (Cln) function in yeast. Cell 66: 1197-1206.
87. Lewin B (1990) Driving the cell cycle: M phase kinase, its partners, and substrates. Cell 61: 743-752.
88. Lin X, Feng XH and Watson JC (1991) Differential accumulation of transcripts encoding protein kinase homologs in greening pea seedlings. In: 3rd Int Congress of Plant Mol Biol, No. 1066.
89. Ling V and Zielinski RE (1989) Cloning of cDNA sequences encoding the calcium-binding protein, calmodulin, from barley (*Hordeum vulgare* L.). Plant Physiol 90: 714-719.
90. Lucantoni A and Polya GM (1987) Activation of wheat embryo calcium-regulated protein kinase by unsaturated fatty acids in the presence and absence of calcium. FEBS Lett 221: 33-36.
91. Ma H, Yanofsky MF and Huang H (1991) Isolation and sequence analysis of TGA1 cDNAs encoding a tomato G protein subunit. Gene 107: 189-195.
92. Ma H, Yanofsky MF and Meyrowitz EM (1990) Molecular cloning and characterization of GPA1, a G protein a subunit gene from *Arabidopsis thaliana*. Proc Natl Acad Sci USA 87: 3821-3825.
93. Magyar Z, Bakó L, Bögre L, Dedeoglu D, Kapros T and Dudits D (1993) Active cdc2 genes and cell cycle phase specific cdc2-related kinase complexes in hormone stimulated alfalfa cells (submitted).
94. Marmé D (1989) The role of calcium and calmodulin in signal transduction. In: Second Messengers in Plant Growth and Development. WF Boss, DJ Morré (eds), pp. 57-80. New York: Alan R. Liss.
95. Martiny-Baron G and Scherer GFE (1989) Phospholipid-stimulated protein kinase in plants. J Biol Chem 264: 18052-18059.
96. McClure BA and Guilfoyle TJ (1987) Characterization of a class of small auxin-inducible polyadenylated RNAs. Plant Mol Biol 9: 611-623.
97. McClure BA, Hagen G, Brown CS, Gee MA and Guilfoyle TJ (1989) Transcription, organization sequence of an auxin-regulated gene cluster in soybean. The Plant Cell 1: 229-239.
98. Minocha SC (1979) Abscisic acid promotion of cell division and DNA synthesis in Jerusalem artichoke tuber tissue cultured *in vitro*. Z. Pflanzenphysiol 92: 327-339.
99. Minshull J, Golsteyn R, Hill CS and Hunt T (1990) The A- and B-type cyclin associated *cdc*2 kinases in *Xenopus* turn on and off at different times in the cell cycle. EMBO J 9: 2865-2875.
100. Minshull J, Pines J, Golsteyn R, Standart N, Mackie S, Colman A, Blow J, Ruderman JV, Wu M and Hunt T (1989) The role of cyclin synthesis, modification and destruction in the control of cell division. J Cell Sci Suppl 12: 77-97.
101. Moreno S and Nurse P (1990) Substrates for p34[cdc2] *in vivo* veritas? Cell 61: 549-551.
102. Morré DJ, Gripshover B, Monroé A and Morré JT (1984) Phosphatidylinositol turnover in isolated soybean membranes stimulated by the synthetic growth hormone 2,4-dichlorophenoxyacetic acid. J Biol Chem 259: 15364-15368.
103. Morse MJ, Satter RL, Crain RC and Coté GG (1989) Signal transduction and phosphatidylinositol turnover in plants. Physiol Plant 76: 118-121.
104. Motokura T, Bloom T, Kim HG, Jüppner H, Ruderman JV, Kronenberg HM and Arnold A (1991) A BCL1-linked candidate oncogene which is rearranged in parathyroid tumors encodes a novel cyclin. Nature 350: 512-518.
105. Murray MG and Key JL (1978) 2,4-dichlorophenoxyacetic acid-enhanced phosphorylation of soybean nuclear proteins. Plant Physiol 61: 190-198.
106. Napier RM and Venis A (1990) Monoclonal antibodies detect an auxin-induced confor-

mational change in the maize auxin-binding protein. Planta 182: 313-318.

107. Ninomya-Tsui J, Yasuda H, Nomoto S, Reed SI and Matsumoto K (1991) Cloning of a human cDNA encoding a *cdc*2-related kinase by complementation of a budding yeast *cdc*28 mutation. Proc Natl Acad Sci USA 88: 1006-1009.

108. Nishi A, Kato K, Takahashi M and Yoskida R (1977) Partial synchronization of carrot cell cultures by auxin deprivation. Physiol Plant 39: 9-12.

109. Nishizuka Y (1986) Studies and perspectives of protein kinase C. Science 233: 305-311.

110. Norbury CJ and Nurse P (1989) Control of higher eukaryote cell cycle by p34^{cdc2} homologues. Biochim Biophys Acta 989: 85-95.

111. Nurse P (1990) Universal control mechanism regulating onset of M-phase. Nature 344: 503-508.

112. Nurse P and Bissett Y (1981) Gene required in G1 for commitment to cell cycle and in G2 for control of mitosis in fission yeast. Nature 292: 558-560.

113. Ohno S, Kawasaki H, Imajoh S, Suzuki K, Inagaki M, Yokokura H, Sakoh T and Hidaka H (1987) Tissue specific expression of three distinct types of rabbit brain protein kinase C. Nature 325: 161-166.

114. Oláh Z and Kiss Z (1986) Occurrence of lipid and phorbol ester activated protein kinase in wheat cells. FEBS Lett 195: 33-37.

115. Oláh Z, Bögre L, Lehel Cs, Faragó A, Seprödi J and Dudits D (1989) The phosphorylation site of Ca^{2+}-dependent protein kinase from alfalfa. Plant Mol Biol 12: 453-461.

116. Palme K (1992) Molecular analysis of plant signaling elements: relevance of eukaryotic signal transduction models. Int Rev Cytol 132: 223-283.

117. Palme K, Hesse T, Moore I, Campos N, Feldwisch J, Garbers C, Hesse F and Schell J (1991) Hormonal modulation of plant growth: the role of auxin perception. Mechanisms of Development 33: 97-106.

118. Piazza GJ (1988) Calmodulin in plants. In: Calcium-Binding Proteins. MP Thompson (ed), pp. 127-143 Boca Raton: FL, CRC Press.

119. Pines J and Hunter T (1989) Isolation of human cyclin cDNA: Evidence for cyclin mRNA and protein regulation in the cell cycle and for interaction with p34^{cdc2}. Cell 58: 833-846.

120. Pines J and Hunter T (1990) p34^{cdc2}: the S and M kinase? New Biol 2: 389-401.

121 Poovaiah BW, Reddy ASN and McFadden JJ (1987) Calcium messenger system; role of protein phosphorylation and inositol bisphospholipids. Plant Physiol 69: 569-573.

122. Prasad PV and Jones AM (1991) Putative receptor for the plant growth hormone auxin identified and characterized by anti-idiotypic antibodies. Proc Natl Acad Sci USA 88: 5479-5483.

123. Prywes RA, Dutta A, Cromlish JA and Roeder RG (1988) Phosphorylation of serum response factor, a factor that binds to the serum response element of the c-fos enhancer. Proc Natl Acad Sci USA 85: 7206-7210.

124. Putnam-Evans C, Harmon AC and Cormier MJ)1990) Purification and characterization of a novel calcium-dependent protein kinase from soybean, Biochemistry 29: 2488-2495.

125. Ralph RK, McCombs PJA, Tener G and Wojicik SJ (1972) Evidence for modification of protein phosphorylation by cytokinins. Biochem J 130: 901-911.

126. Ranjeva R and Boudet AM (1987) Phosphorylation of proteins in plants: regulatory effects and potential involvement in stimulus/response coupling. Ann Rev Plant Physiol 38: 73-93.

127. Ranjeva R, Carrasco A and Boudet AM (1988) Inositol trisphosphate stimulates the release of calcium from intact vacuoles isolated from *Acer* cells, FEBS Lett 230: 137-141.

128. Reddy ASN, Chengappa S and Poovaiah BW (1987) Auxin-regulated changes in protein phosphorylation in pea epicotyls. Biochem Biophys Res Commun 144: 944-950.

129. Reed SI (1991) G1-specific cyclins: in search of an S-phase-promoting factor. Trends in Genetics 7: 95-99.

130. Rincon M and Boss WF (1987) Myo-inositol trisphosphate mobilizes calcium from fusogenic carrot (*Daucus carrota* L.) protoplasts. Plant Physiol 83: 395-398.

131. Roberts DM, Lukas TJ and Watterson DM (1986) Structure, function and mechanism of

action and calmodulin. Crit Rev Plant Sci 4: 311-339.

132. Sanders D, Hansen UP, Slayman CL (1981) Role of the plasma membrane proton pump in pH regulation in non-animals cells. Proc Natl Acad Sci USA 78: 5903-5907.

133. Schafer A, Bygrave F, Matzenauer S and Marmé D (1985) Identification of a calcium-a phospholipid-dependent protein kinase in plant tissue. FEBS Lett 187: 25-28.

134. Schafer W and Kahl G (1981) Auxin-induced changes in chromosomal protein phosphorylation in wounded potato tuber parenchyma. Plant Mol Biol 1: 5-17.

135. Scherer GFE, André B and Martiny-Baron G (1990) Hormone-activated phospholipase A2 and lysophospholipid-activated protein kinase: a new signal transduction chain and a new second messenger system in plants? Current Topics Plant Biochem Physiol 9: 190-218.

136. Schumaker KS and Sze H (1987) Inositol 1,4,5-trisphosphate releases Ca^{2+} from vacuolar membrane vesicles of oat roots. J Biol Chem 262: 3944-3946.

137. Schroeder JI and Thuleau P (1991) Ca^{2+} channels in higher plant cells. Plant Cell 3: 555-559.

138. Smith H (1990) Signal perception, differential expression within multigene families and the molecular basis of phenotypic plasticity. Plant, Cell and Env 13: 585-594.

139. Solomon MJ, Glotzer M, Lee TH, Philippe M and Kirschner MW (1990) Cyclin activation of p34[cdc2]. Cell 63: 1013-1024.

140. Suen KL and Choi JH (1991) Isolation and sequence analysis of a cDNA clone for a carrot calcium-dependent protein kinase: homology to calcium/calmodulin-dependent protein kinases and to calmodulin. Plant Mol Biol 17: 581-590.

141. Surana U, Robitsch H, Price C, Schuster T, Fitch I, Futcher AB and Nasmyth K (1991) The role of cdc28 and cyclins during mitosis in the budding yeast S. cerevisiae. Cell 65: 145-161.

142. Sussmann MR and Harper JF (1989) Molecular biology of the plasma membrane of higher plants. Plant Cell 1: 953-960.

143. Takahashi Y, Kuroda H, Tanaka T, Machida Y, Takebe I and Nagata T (1989) Isolation of an auxin-regulated gene cDNA expressed during the transition from G0 to S-phase in tobacco mesophyll protoplasts. Proc Natl Acad Sci USA 86: 9279-9283.

144. Theologis A, Huynh TV and Davis RW (1985) Rapid induction of specific mRNAs by auxin in pea epicotyl tissue. J Mol Biol 183: 53-68.

145. Thompson MP, Piazza GG, Brower DP, Jr. Farrell HM (1989) Purification and characterization of calmodulins from Papaver somniferum and Euphorbia lathyris. Plant Physiol 89:501-505.

146. Trewavas A and Gilroy S (1991) Signal transduction in plant cells. Trend in Genetics 7: 356-361.

147. Trewavas AJ and Jennings DH (1986) Introduction. In: Plasticity in Plants, Symposia of the Society for Experimental Biology. Number 40: 1-4.

148. Tsai LH, Harlow E and Meyerson M (1991) Isolation of the human cdk2 gene that encodes the cyclin A- and adenovirus E1A-associated p33 kinase. Nature 353: 174-177.

149. Van 't Hof J and Kovács CJ (1972) Mitotic cycle regulation in the meristem of cultured roots. The principal control point hypothesis. Ad Exp Med Biol 18: 15-30.

150. Veluthambi K and Poovaiah BW (1986) in vitro and in vivo protein phosphorylation in Avena sativa L coleoptiles. Plant Physiol 81: 836-841.

151. Verde F, Labbé JC, Dorée M and Karsenti E (1990) Regulation of microtubule dynamics by cdc2 protein kinase in cell-free extracts of Xenopus eggs. Nature 343: 233-238.

152. Wang H, Saleem M, Fowke LC and Cutler AJ (1991) DNA synthesis in maize protoplasts in relation to source tissue differentiation. J Plant Physiol 138: 200-203.

153. Waterborg JH, Harrington RE and Winicov I (1989) Differential histone acetylation in alfalfa (Medicago sativa) due to growth in NaCl. Plant Physiol 90: 237-245.

154. Wenzler H and Meins F (1986) Mapping of the maize leaf capable of proliferation in culture. Protoplasma 131: 103-105.

155. Wernicke W and Milkovits L (1987) Effect of auxin on the mitotic cell cycle in cultured leaf segments at different stages of development in wheat. Physiol Plant 69: 16-22.

156. Wick SM, Muto S and Duniec J (1985) Double immunofluorescence labeling of calmodulin

and tubulin in dividing plant cells. Protoplasma 126: 198-206.

157. Yamamoto KK, Gonzalez GA, Biggs II WH and Montminy MR (1988) Phosphorylation-induced binding and transcriptional efficacy of nuclear factor CREB. Nature 334: 494-498.

158. Yasuda T, Yajima Y and Yamada Y (1974) Induction of DNA synthesis and callus formation from tuber tissue of Jerusalem artichoke by 2,4-dichlorophenoxyacetic acid. Plant Cell Physiol 15: 321-329.

159. Yeoman MM and Aitchison PA (1973) Growth patterns in tissue (callus) cultures. In: Plant Tissue and Cell Culture. HE Street (ed), pp. 240-268, Oxford: Blackwell Scientific Publication.

160. Zbell B and Walter-Back C (1988) Signal transduction of auxin on isolated plant cell membranes: indications for a rapid polyphosphoinositide response stimulated by indoleacetic acid. J Plant Physiol 133: 353-360.

161. Zocchi G (1985) Phosphorylation-dephosphorylation of membrane proteins controls the microsomal H^+-ATPase activity of corn roots. Plant Sci 40: 153-159.

10. Auxin-mediated activation of DNA synthesis via par genes in tobacco mesophyll protoplasts

TOSHIYUKI NAGATA and YOHSUKE TAKAHASHI

Abstract

Tobacco mesophyll protoplasts undergo a transition from G0- to S-phase upon culture on an appropriate medium containing auxin and cytokinin. In this process auxin plays a crucial role in inducing meristematic activity of differentiated and non-dividing cells. When we searched for genes which were induced upon the addition of auxin to the medium, three auxin-regulated genes; *par*A, *par*B and *par*C were isolated. Northern hybridization revealed that these genes were in fact expressed during the progression from G0- to S-phase. The predicted gene product of *par*A is proposed to play a role in transcriptional regulation, while that of *par*B encodes glutathione *S*-transferase (GST). The GST is the first example, in which an auxin-regulated gene is ascribed to a specific enzymatic activity. In the promoter of *par*A gene, an auxin-responsive region has been identified. The significance of the expression of these genes and their gene products is discussed in relation to the molecular mode of auxin action.

1. Introduction

Auxin was discovered more than 60 years ago as the first plant hormone and has been reported to influence the most widespread aspect of the growth, development and differentiation of plants [21]. It is well documented that auxin is necessary for the culture of plant tissues in *in vitro* conditions (see D Dudits *et al.*, this volume) although its effect on cell expansion has been studied more intensively. However, the molecular mechanism of the action of auxin either on cell division or on cell expansion has not been elucidated. Hence, it is a challenge to understand the role of auxin on the induction of meristematic activity in plant cells at the molecular level.

The culture of tobacco mesophyll protoplasts has been established since 1970 [8, 9]. One of the characteristic features of this experimental system is that the tobacco mesophyll protoplasts divide semi-synchronously at the early stage of culture, and form colonies at high frequency when they are cultured in an appropriate condition. The addition of two plant hormones, auxin and cytokinin, is required absolutely for the induction of meristematic activity, but the effect of auxin on this process is much more pronounced. Prior to the

J.C. Ormrod and D. Francis (eds.), Molecular and Cell Biology of the Plant Cell Cycle, 133–141.
© 1993 *Kluwer Academic Publishers. Printed in the Netherlands.*

activation of cell division, non-dividing mesophyll cells are considered to be in G0 of the cell cycle. Thus, in order to understand the role of auxin during the transition from G0- to S-phase, we looked for genes which could be induced during this transition in cultured mesophyll cells. Thus far we have isolated three auxin-regulated genes from the early stage of the cultured tobacco mesophyll protoplasts [14, 16, 17]. In this paper we will discuss the possible function and significance of these genes in inducing meristematic activity in plant cells.

2. Materials and methods

Preparation and culture of tobacco mesophyll protoplasts was carried out as described before [14]. Most of the molecular biological procedures were conducted as before [14]. The auxin-responsiveness of *par*A promoter was determined by the assay of β-glucuronidase (GUS) after the delivery of chimaeric plasmids into tobacco mesophyll protoplasts by electroporation [18]. DNA synthesis of the cultured tobacco mesophyll protoplasts was assessed by the incorporation of ^{3}H-thymidine into the trichloroacetic acid (TCA)-insoluble fraction after the incubation of the protoplasts with the radioactive nucleoside (100 μCi ml-^{1}) for 30 min.

3. Results and discussion

3.1. Requirement of plant hormones for the induction of DNA synthesis in tobacco mesophyll protoplasts

The method of culturing tobacco mesophyll protoplasts has been established for 20 years [8, 9]. These previous studies revealed that the addition of suitable combinations of auxin and cytokinin to an appropriate medium affects the fate of cultured protoplasts most significantly. Furthermore, we showed in 1970 that DNA synthesis and subsequent limited cell division can be induced in a relatively simple medium [7]. Hence, we chose the simple medium for the culture of the protoplasts, and avoided uncontrolled interaction which could occur in a more complex medium. The latter medium is suitable for obtaining sustained cell divisions [9].

Combinations of auxin and cytokinin were necessary for the induction of DNA synthesis which did not occur if either auxin or cytokinin was deleted from the medium (Fig. 1). DNA synthesis was observed only in the presence of auxin and cytokinin at 20-24 h after the start of culture. Accordingly, cell division was observed in the presence of auxin and cytokinin after 24 h of culture. However, a limited, but significant level of the incorporation of ^{3}H-thymidine into the acid insoluble fraction was observed in the medium lacking cytokinin at 72 h of culture. Although we have not confirmed whether this

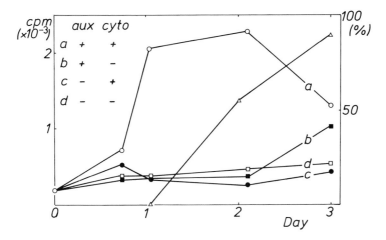

Fig. 1. Dependency of the initiation of DNA synthesis of tobacco mesophyll protoplasts on auxin and cytokinin. Tobacco mesophyll protoplasts were cultured in the medium supplemented with auxin [2,4-D, 1 mg/l] and cytokinin [BAP, 1 mg/l] (a), with 2,4-D (b), with BAP (c) and without plant hormones (d). The triangles represent the time course of divided protoplasts in the medium supplemented with 2,4-D and BAP. Without either of plant hormones, no cell division was observed.

incorporation of ³H-thymidine could reflect the actual DNA synthesis, cytokinin habituation might be involved in this case as has been noticed in the culture of tobacco pith tissue by Meins *et al.* [6]. Thus, we thought that some specific synthesis of macromolecules should be induced before the onset of DNA synthesis. Hence, we looked for specific transcripts which could be preferentially induced by auxin. When using a cDNA probe which had been prepared from mRNA isolated from the protoplasts cultured for 24 h in the presence of auxin, we conducted differential screening. Three auxin-regulated genes have been isolated, which we named *par* (protoplast *a*uxin-regulated) genes. They are *par*A [14], *par*B [16] and *par*C [17]. Furthermore, from this cDNA library another gene *C*-7 was isolated which has homology to *par*A and *par*C, but did not show any response to auxin treatment [17].

*3.2. Time course of the expression of par*A, *par*B *and par*C *genes*

Northern hybridization experiments using *par*A and *par*B as probes revealed that the expression of these genes was detected as early as 20 min after the addition of auxin, reached a maximum after 4-6 h. This level of expression began to decline after 1 day, when active DNA synthesis was observed. After 2 days, when active cell division was observed, the expression of both genes was almost suppressed [15, 16]. Although some residual level of expression of *par*C was observed at the start of culture, the time course of the expression of *par*C was essentially similar to that of *par*A and *par*B. Thus, the expression of *par* genes was observed during the cell cycle progression of tobacco

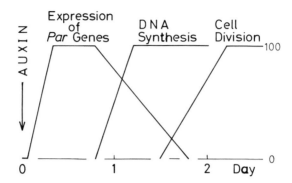

Fig. 2. Schematic, illustration of the time course of the addition of auxin, the expression of *par*A, *par*B and *par*C, DNA synthesis and cell division.

mesophyll protoplasts from G0 to S-phase. This sequence is schematically illustrated in Fig. 2. The next question was what is the function of these gene products and what roles do they play during the cell cycle progression from G0- to S-phase?

3.3. ParA gene

The predicted amino acid sequence of the *par*A cDNA showed that the gene product consists of 220 amino acids. The homology search in the GenBank revealed that *par*A has homology to an auxin-regulated and heat-shock responsive gene, *Gmhsp 26-A* [3] from soybean as well as to the stringent starvation protein (SSP) of *Escherichia coli* [12]. The former does not offer any clues about the function of *par*A, but the latter sheds some light, since SSP binds equimolar to an apoprotein of RNA polymerase [4], and may regulate gene expression at the level of transcription. *Par*A could have such a function in transcriptional regulation in plants given the structural homology between the RNA polymerase in prokaryotes and RNA polymerase II in eukaryotes [15]. If this assumption is correct, the *par*A gene product should be localized in the nucleus for the fulfillment of its function and be associated with RNA polymerase II. In fact, the staining of tobacco mesophyll protoplasts cultured for 6 h with a polyclonal antibody raised against the *par*A gene product revealed that the *par*A gene product was confined to the nucleus, while the freshly prepared protoplasts were not stained with this antibody (Y Takahashi, S Hasezawa and T Nagata, unpublished). This result indicated that the *par*A gene product would have been transported to the nucleus by some specific mechanism.

3.4. ParB gene

The sequence determination of *par*B cDNA revealed that the *par*B gene encodes a protein consisting of 213 amino acids and the predicted gene product has homology to a maize glutathione *S*-transferase III (GST; RX: glutathione R-transferase, EC 2.5.1.18). Furthermore, it has homology to a rat lysophospholipase as well as other GSTs including placental ones. To examine whether the gene product of *par*B is in fact GST, we inserted the coding sequence of *par*B into a bacterial expression vector, pKK233-2. When the *E. coli* harbouring pKK233-2 was induced with ITPG, the bacteria produced fusion proteins which showed high activity of GST according to a standard assay of GST using 1-chloro-2,4-dinitrobenzene as a substrate [16]. This is the first example, in which an auxin-regulated gene has been ascribed to a specific function. Subsequently it would be intriguing to discern the role of this GST during the cell cycle progression from G0- to S-phase in tobacco mesophyll protoplasts. At this moment there are two possibilities: one, the *par*B gene product plays a role in detoxification of electrophilic xenobiotics in catalysing the conjugation with glutathione, as has been widely demonstrated in plants and animals [2, 10]. Two, the *par*B gene product could have a similar function to that of placental GSTs, some of which are reported to be induced specifically at an early stage of chemical hepatocarcinogenesis in rat; its causal relationship is almost 100% [11]. Moreover experimentally-induced expression of the activated *ras* gene results in concomitant expression of a placental GST in cultured rat liver cells in association with carcinogenesis [5]. The fact that the GST binds with indole-3-acetic acid (K Palme, J Schell, personal communication) may be linked with its role. The subcellular localization of *par*B gene products was found to be in cytoplasm and plasma membranes when the cells were examined under fluorescence microcopy after staining with the antibody against *par*B gene products (S Hasezawa, Y Takahashi, T Nagata, unpublished).

3.5. ParC gene and its related C-7 gene [17]

The sequence of *par*C cDNA predicts a gene product consisting of 221 amino acids. Because the *par*C and *par*A gene products showed homology, we looked for other *par*A-related genes in the cDNA library from the cultured tobacco mesophyll protoplasts. Subsequently a gene *C*-7 was isolated whose gene product is predicted to consist of 219 amino acids. Although *C*-7 has homology to *par*A and *par*C, *C*-7 did not respond to auxin treatment. The expression of *C*-7 was observed predominantly in differentiated organs (leaves and stems), while the expression of *par*A, *par*B and *par*C was observed primarily in root tips.

Since it became apparent that several genes belong to a gene family of *par*A, we searched for other related genes in the literature and in the GenBank data base. This survey revealed that *prp*1 gene from potato [19], two genes *103* and *107* from cultured cells of tobacco [20] as well as *Gmhsp 26*-A from soybean belong to this gene family (Table 1). All of these genes have homology to SSP. *Gmhsp 26*-A gene responds to auxin and other stresses [3]. *Prp*1 gene was induced in potato upon infection with a pathogenic fungus, *Phytophthora*

Table 1. Characterisation of *par*-A related genes

	*par*A	*par*C	C-7	103	107	*prpl*	*Gmhsp26*-A
source of cDNA	tobacco mesophyll	tobacco mesophyll	tobacco mesophyll	tobacco cell	tobacco cell	potato leaves	soybean seedling
inducer	auxin	auxin	–	auxin	auxin	pathogen	stress
tissue-specific expression	root	all organs	all organs	root tip	–	ND	ND
ATTA sequence	3	1	4	0	1	1	1
ref.	Takahashi *et al.* 1989	this work		vanderZaal *et al.* 1991		Taylor *et al.* 1990	Czarnecka *et al.* 1988

infestans [19], while *103* and *107* are auxin regulated genes [20]. Thus, a common function of the members of *par*A gene family could be transcriptional regulation because of the homology to SSP. Moreover, each gene could regulate gene expression responding to different external stimuli such as auxins, fungal elicitors and various physical and chemical stresses including heat shock and heavy metal ions. In this context it should be noted that the examination of the 3-D structures of the products of *par*A-related genes have consistently shown the domains of acidic activators. Although any molecular biological proof still awaits, all of the gene products of *par*A-related genes could play a role in transcriptional regulation responding to various external and internal signals.

3.6. Transcriptional regulation of parA gene

Transcriptional regulation on the auxin-regulated *par*A gene has been studied [18]. We isolated the *par*A gene from a tobacco genomic library using *par*A cDNA as a probe. In order to examine the auxin responsiveness of the *par*A promoter, a chimaeric gene was constructed in which a 0.8 kb *Mae* II fragment of the *par*A promoter was fused to the coding region of the *E. coli* β-glucuronidase (GUS) gene. When the chimaeric DNA was introduced into tobacco mesophyll protoplasts by electroporation, auxin-responsiveness was observed in this region. Subsequently, construction of a series of deletion mutants and the delivery of these genes into protoplasts revealed that auxin responsiveness is confined to a 111 bp region in the promoter. A more recent trial showed that there is a protein which binds preferentially to specific sequences within the 111 bp repeat. This sequence could be an auxin-responsive element and the protein could be a *trans*-acting factor to auxin (M Kusaba, Y Takahashi, T Nagata unpublished).

In conjunction with the studies on auxin-regulated genes, the relationship between the expression of these genes and auxin receptors was examined. An auxin-binding protein (ABP) of 22 kDa was identified in the membrane fraction of tobacco mesophyll protoplasts, which was immunologically cross-reacted with a well-characterized maize ABP having a molecular mass of 21 kDa [13]. The binding of ABP with the surface of the plasmalemma in tobacco mesophyll protoplasts suggests that it plays a role in signal perception [1]. Auxin signals may be perceived on the surface of tobacco mesophyll protoplasts (see D Dudits *et al.*, this volume) resulting in the expression of several genes such as *par*A, *par*B and *par*C. These gene products would then activate the differentiated and non-dividing tobacco mesophyll protoplasts into the cell cycle. Thus, we propose for the first time, the chain from auxin application to gene expression through to the initiation of meristematic activity in tobacco mesophyll protoplasts.

3.7. Implication of the expression of parA, parB and parC genes in inducing meristematic activity in tobacco mesophyll protoplasts

When we interpret the profile of the expression of the *par*A, *par*B and *par*C genes, the product of the most abundantly expressed *par*A gene could play a key role in transcriptional regulation, resulting in the cell cycle progression from G0- to S-phase. In this consideration, description of *par*A and *par*C is only represented by that of *par*A, as the structure and function of *par*C is similar to that of *par*A. Furthermore, since during the cell cycle progression in highly synchronized tobacco BY-2 cells [7], the level of expression of *par*A was low and showed no significant changes during the four cell cycle stages, the gene product of *par*A would play a pivotal role in forcing the entry of non-dividing differentiated cells into the cell cycle, but not affecting the duration of the cell cycle. However, it would be possible that even in the normal cell cycle a very low level of expression of *par*A is required, although such a possibility has not been examined yet. In this context perhaps the genes *par*A, *par*B and *par*C do not have any direct relation to cell cycle genes such as *cdc*2 which have been shown to be involved in the cell cycle progression in eukaryotic cells. The fact that the *par*A-related genes, responding to various external and internal signals, could represent the unique characteristics of plant cells that could be related to their totipotency.

On the other hand, regarding the significance of the *par*B gene expression, we have not concluded yet whether GSTs, as the gene product of *par*B, have a role in detoxification or in neoplasmic transformation. As the time course expression of *par*B coincided with that of *par*A, it should have some significant role. Thus, we suppose that the *par*B gene product could play a role primarily in the cytoplasm and plasma membranes in contrast to the role of the *par*A gene product which is located in the nucleus. In this context, the observation by Palme and Schell (personal communication) that one auxin-binding protein in the plasma membrane fraction of *Arabidopsis thaliana* was found to be structurally almost similar to the GST (the gene product of *par*B) suggests an

140

interesting regulatory function of the *par*B gene product. Moreover, the expression of *par*B was regulated by auxin and, subsequently, its gene product could be regulated by binding with auxin.

However, several points remain to be resolved. It is probable that using differential screening only abundantly-expressed genes are isolated. Hence, a further search for less-abundantly expressed genes is necessary and may involve the use of specific mutants. Furthermore, the role of cytokinin in inducing DNA synthesis in tobacco mesophyll protoplasts has not been examined in detail and we do not know what kinds of genes could be involved. It will also be necessary to examine the relationship of the auxin-regulated genes from tobacco mesophyll protoplasts to those from auxin-regulated genes from elongating tissues.

Acknowledgements

This study was supported in part by grants from the Ministry of Education, Culture and Science of Japan and by grants from the Ministry of Agriculture, Forestry and Fisheries of Japan to T.N.

References

1. Barbier-Brygoo H, Ephritikhine G, Klämbt D, Ghislain, M and Guern J (1989) Functional evidence for an auxin receptor at the plasmalemma of tobacco mesophyll protoplasts. Proc Natl Acad Sci USA 86: 891-891.
2. Chasseaud LF (1979) The role of glutathione and glutathione *S*-transferases in the metabolism of chemical carcinogenesis and other electrophilic agents. Adv Cancer Res 29: 175-274.
3. Czarnecka E, Nagao RT, Key JE and Gurley WB (1988) Characterization of *Gmhsp 26*-A, a stress gene encoding a divergent heat shock protein of soybean; heavy-metal-induced inhibition of intron processing. Mol Cell Biol 8: 1113-1122.
4. Ishihama A and Saitoh T (1979) Subunits of RNA polymerase in function and structure. IX. Regulation of RNA polymerase activity by stringent starvation protein (SSP). J Mol Biol 129: 517-530.
5. Li U, Seyama T, Godwin AK, Winokur TS, Lebovitz RM and Lieberman MW (1988) M*Tras*T24, a metallothionein-*ras* fusion gene, modulates expression in cultured rat liver cells of two genes associated with *in vivo* liver cancer. Proc Natl Acad Sci USA 84: 344-348.
6. Meins Jr F, Lutz J and Foster R (1980) Factors influencing the incidence of habituation for cytokinin of tobacco pith tissue in culture. Planta 150: 264-268.
7. Nagata T, Nemoto Y and Hasezawa S (1992) Tobacco BY-2 cell line as the 'HeLa' cell in the cell biology of higher plants. Int Rev Cytol 132: 1-30.
8. Nagata T and Takebe I (1970) Cell wall regeneration and cell division in isolated tobacco mesophyll protoplasts. Planta 92: 301-308.
9. Nagata T and Takebe I (1971) Plating of isolated tobacco mesophyll protoplasts on agar medium. Plants 99: 12-20.
10. Rushmore TH, King RG, Paulson KE and Pickett CB (1990) Regulation of glutathione *S*-transferase Ya subunit gene expression: Identification of a unique xenobiotic-responsive element controlling inducible expression by planar aromatic compounds. Proc Natl Acad Sci USA 87: 3826-3830.

11. Satoh K, Kitahara, Soma Y, Inaba Y, Hatayama I and Sato K (1985) Purification, induction, and distribution of placental glutathione transferase: A new marker enzyme for preneoplastic cells in the rat chemical hepatocarcinogenesis. Proc Natl Acad Sci USA 82: 3964-3968.

12. Serizawa H and Fukuda R (1987) Structure of the gene for the stringent starvation protein of *Escherichia coli*. Nucl Acids Res 15: 1153-1163.

13. Shimomura S, Liu W, Nagata T and Futai M (1992) Comparison of membrane auxin-binding proteins from maize and tobacco (submitted).

14. Takahashi Y, Kuroda H, Tanaka T, Machida Y, Takebe I and Nagata T (1989) Isolation of an auxin-regulated gene cDNA expressed during the transition from G0 to S phase in tobacco mesophyll protoplasts. Proc Natl Aca Sci USA 86: 9279-9283.

15. Takahashi Y, Kusaba M, Hiraoka Y and Nagata T (1991) Characterization of the auxin-regulated *par* gene from tobacco mesophyll protoplasts. Plant Journal 1: 327-332.

16. Takahashi Y and Nagata T (1992) *par*B: An auxin-regulated gene encoding glutathione *S*-transferase. Proc Natl Acad Sci USA 89: 56-59.

17. Takahashi Y and Nagata T (1992) Differential expression of auxin-regulated *par*C gene and its related gene *C*-7 from tobacco mesophyll protoplasts to external stimuli and in tissues. Plant Cell Physiol 33: 779-787.

18. Takahashi Y, Niwa Y, Machida Y and Nagata T (1990) Location of the cis-acting auxin-responsive region in the promoter of the *par* gene from tobacco mesophyll protoplasts. Proc Natl Acad Sci USA 87: 8013-8016.

19. Taylor JL, Fritzemerier KH, Hauser I, Kombrink E, Rohwer F, Schroder M, Strittmatter G and Hahlbrock K (1990) Structural analysis and activation by fungal infection of a gene encoding a pathogenesis-related protein in potato. Mol Plant-Microbe Int 3: 72-77.

20. Van der Zaal EJ, Droog FNJ, Boot CJM, Hengens LAM, Hoge JHC, Schilperoort RA and Libbenga KR (1991) Promoters of auxin-induced genes from tobacco can lead to auxin-inducible and root tip-specific expression. Plant Mol Biol 16:983-998.

21. Wareing PF, Phillips IDJ (1981) Growth and Differentiation in Plants. 3rd Ed. Oxford, Pergamon.

11. Polyamines and protein modification during the cell cycle

S. DEL DUCA and D. SERAFINI-FRACASSINI

Abstract

The behaviour of polyamines, proteins and other physiological parameters during the synchronous cell cycle following the break of dormancy of tubers of *Helianthus tuberosus* L. is presented. Polyamines seem to be relevant in this event. These growth substances may act in free or bound form; the covalently bound polyamines were analyzed both *in vivo* and in cell-free extracts which allows the determination of transglutaminase activity, the enzyme that links polyamines to proteins. This post-translational modification of proteins might have a structural or regulatory role. Free polyamine levels exhibit a bimodal fluctuation which peaks during the G1/S and the division phases of the cell cycle. This seems to be a common feature of plant and animal cycling cells. The level of conjugated polyamines, whose formation is catalyzed by a *de novo* synthesized transglutaminase, increases during the cell cycle showing a sigmoidal trend. The protein pattern on gel electrophoresis changes during the cell cycle showing an increase in the amount of high molecular mass bands.

1. Polyamines

The growth of all living organisms requires polyamines (PA) and rates of growth are proportional to the amount of these aliphatic cations. The structural characteristic of PA is to have protonated terminal aminic, and in the higher PA, iminic, groups on an aliphatic backbone. Therefore, they behave as organic cations. Moreover they are linear and flexible molecules (Fig. 1). The principal precursors of putrescine (PU) are ornithine and arginine; the addition of aminopropyl groups to PU gives rise to spermidine (SD) and to spermine (SM) (Fig. 2). Specific inhibitors can block several steps in their biosynthetic pathway (see the zig-zag arrows in Fig. 2). In the cell, PA can be found as free molecules or as a bound form linked by hydrogen, ionic and also covalent linkages to macromolecules or to other cellular components. They can be linked covalently to glutamyl residues of proteins by the enzyme, transglutaminase (TGase) and thus are TCA-insoluble, forming mono- and bis γ-glutamyl derivatives (Fig. 2) [14, 24]. In the latter, the length of the polyamine molecule determines the length of the bridges. Other covalent linkages are formed in some plant families with hydroxycinnamic acids [42] and such bound PA are TCA-soluble.

J.C. Ormrod and D. Francis (eds.), Molecular and Cell Biology of the Plant Cell Cycle, 143–156.

144

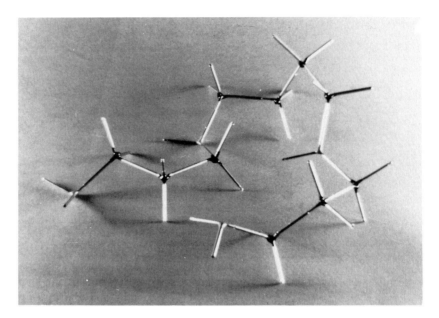

Fig. 1. Model of a putrescine (1,4-diaminobutane) molecule.

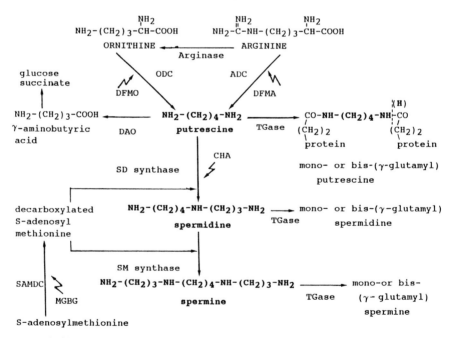

Fig. 2. The biosynthetic pathway and protein linking reaction of polyamines. Abbreviations: ADC = arginine decarboxylase, ARG = arginine, DAO = diamineoxidase, DFMA = difluoromethylarginine, DFMO = difluoromethylornithine, ODC = ornithine decarboxylase, PU = putrescine, SAMDC = S-adenosylmethionine decarboxylase, SD = spermidine, SM = spermine, TGase = transglutaminase.

As with free PA [3, 15], bound PA can also be related to growth. For example, in thin-layer explants of *Nicotiana tabacum* L., both of them peak at the onset of cell division [45]. Data on bound PA during the cell cycle are very scanty [21, 29]. Among the bound TCA-insoluble PA, whose role is unclear, we turned our attention to the very stable conjugated PA, (i.e. covalently-bound to proteins by TGase) which is detected by acidic hydrolysis of the TCA-precipitate. TGase is particularly well known in animals, whose substrates are structural proteins [14]. This enzyme is also present in plants where its function remains to be clarified [9, 16, 19, 28, 37, 38, 41].

To understand the role at the molecular level of free and of bound PA in the cell cycle, we analyzed their pattern *in vivo* (by culturing tuber explants in sterile growth medium also containing labelled PU) and determined the activity of TGase in cell free extracts prepared during the different phases of the cell cycle of *H. tuberosus* [12, 35, 36]. Peptide-PA conjugates (TCA-soluble) were also found in *H. tuberosus,* whereas hydroxycinnamoyl-PA were not detectable.

2. The *Helianthus tuberosus* system

The plants provide a homogeneous tissue: the medullary parenchyma of the tuber. These cells, arrested in G0-phase contain very small amounts of growth

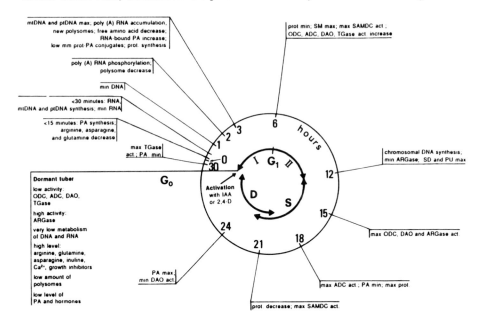

Fig. 3. Main physiological events in parenchyma cells of *H. tuberosus* tuber explants during dormancy and the first cell cycle induced by culturing explants in Bonner and Addicott medium [6] supplemented with 10 μM 2,4-dichlorophenoxyacetic acid (2,4-D) or 2 μM indol-3-acetic acid (IAA). Abbreviations: mtDNA = mitochondria DNA, ptDNA = plastid DNA.

146

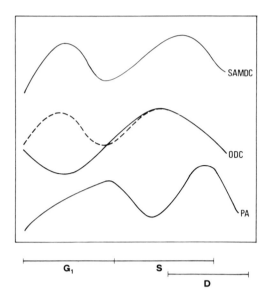

Fig. 4. Relative PA content, ODC and SAMDC activity during the cell cycle of *H. tuberosus, N. tabacum* and animal cells. Dotted lines represent the ODC activity during the cell cycle of some animal cells. Schematic representations redrawn from [17, 18, 19, 36].

substances but a considerable amount of inhibitors (Fig. 3); PA are also present but are in insufficient amounts to sustain growth (Fig. 4) [1]. All the biosynthetic enzymes as well as diamine oxidase (DAO) activities are low or practically absent, while arginase is active (Figs. 2, 3).

Protein content and composition during tuber dormancy are shown in Figs. 5 and 6. Several bands appear particularly evident and their densitometric profiles are reported in Fig. 7. The labelled conjugates formed by incubating [3H] PU with the cell-free extract from dormant tubers are shown in Fig. 6. These conjugates either have a very high molecular mass or are insoluble. Thus, they cannot enter the stacking gel when analyzed by SDS-PAGE [38].

The TGase binding reaction has two components: one is destroyed by boiling (Ts TGase), the other is temperature-insensitive (Ti TGase) [38]. In Fig. 5 the total activity of the crude homogenate was, almost exclusively, the result of Ti activity. This binding might be of a non-enzymatic nature and the complexes formed show a very faint radioactivity on SDS-PAGE autoradiography, suggesting that these PA were not covalently bound. The Ts TGase, whose activity is low during dormancy, possesses catalytic properties similar to those observed in animal TGase [38].

Dormancy can be broken by excising slices of tuber parenchyma and by treating them with IAA, 2,4-D or PA added to the sterile growth medium [5]. The cells abruptly change their function and this allows one to identify the metabolic and morphological parameters involved in this switch.

In the agar-based growth medium, the cells begin to divide synchronously

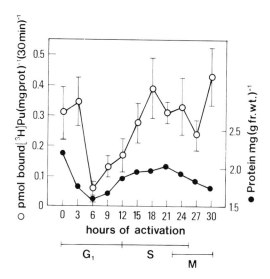

Fig. 5. Mean protein content ± S.E. (●) and mean TGase activity (○) during the first cell cycle of *H. tuberosus* explants cultured as reported in Fig. 3. Protein content was determined by the Bradford method [7]. TGase activity was determined by 10 μCi [³H]PU incorporation in 500 μg protein-containing cell-free extracts from cultured explants. These data result from three replicates.

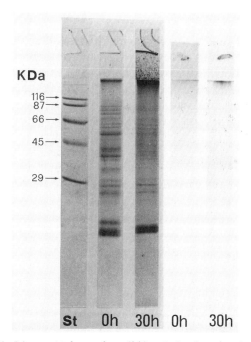

Fig. 6. SDS-PAGE [2] of dormant tuber explants (0 h) and of activated explants whose cells are in division phase (30 h) and relative autoradiography of their cell-free extracts tested for TGase assay after incubation with [³H] PU, as reported in Fig. 5.

148

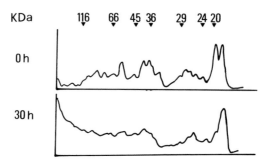

Fig. 7. Densitometric pattern of Coomassie-stainable protein bands analyzed by SDS-PAGE [22] from dormant cells (0 h) and explants cultured as reported in Fig. 3, whose cells are in division phase (30 h). Reference molecular masses are indicated by arrows.

and after 20 days the explants are fully differentiated and present the typical anatomy of the intact tubers. With this *in vitro* system it was demonstrated, for the first time in plants, that PA can act as growth substances [5]. This growth is due mainly to cell division and not cell enlargement and for this reason the involvement of PA in the cell cycle was studied.

3. The cell cycle

By placing slices of tuber parenchyma in liquid medium it is possible to induce the start of a synchronous cell cycle. G1-phase is rather long (12 h) and can be divided into two parts. In the early part (from 0 to 6 h), the cell is involved, presumably, in dedifferentiation and in acquiring the new meristematic characteristics, fully expressed in the second part (from 6 to 12 h). This two-stage hypothesis is consistent with subsequent cell cycles which have shorter G1-phases. Autoradiographic studies have indicated that the S-phase of the first cell cycle begins after the 12th hour and ends around the 25th hour. The G2-phase is virtually zero since division starts in some cells, while other cells are still in S-phase [36]. The cell cycle duration under these conditions is about 30 h.

The morphological aspects of the cycling cells have been described previously [12, 35]. At about 20 h an increasing number of cells contain two less densely stained nuclei, with several nucleoli; the nuclei are separated by a thin wall apparently growing centripetally [35]. Although mitoses have been reported to occur [36], studies performed using colchicine did not reveal the expected accumulation of metaphase nuclei. Moreover, in the first cell cycle many of the nuclei divide amitotically, i.e. without any chromosomal condensation and formation of a mitotic spindle; in the subsequent cell cycles, normal mitoses become detectable [35]. Cell divisions cease around 27–30 h [35, 36].

4. G1-phase

PA biosynthesis is probably one of the earliest metabolic events induced by activation since an increase in their content can already be measured at 15 min [36]. The level attained after 1 h is already sufficient to sustain protein synthesis. Free PA continue to increase throughout the G1-phase (Fig. 4).

DNA, RNA, proteins and polysomes have a high turnover because they are very rapidly degraded but all are also newly synthesized (Figs. 3, 5) [36]. PA can bind also to RNA. Spermine (SM) maintains phe-tRNA in its biologically active conformation and enhances the efficiency and fidelity of translation [31]. The SM:phe tRNA ratio is similar to the ratio found *in vivo* in extracts from different plants [2]. In *H. tuberosus*, the number of PA molecules bound to the different RNAs increases in early G1-phase [40]. Ribosomes extracted from dormant tuber cells, contrary to those extracted from dividing cells, required the addition of PA and Mg^{2+} to be active in translation. With regard to protein synthesis, both labelled methionine and leucine were incorporated

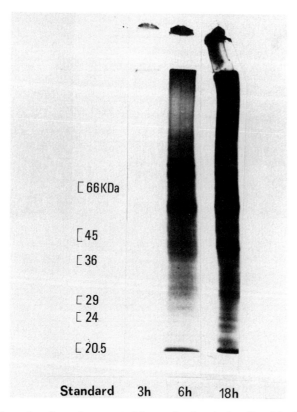

Fig. 8. Autoradiography of proteins extracted from cells of explants cultured for 3 h, 6 h or 18 h, as reported in Fig. 3. The medium also contained 15 μCi [14C] leucine per 5 g of fresh tissue. Proteins were determined by the Bradford method [7].

150

Table 1. Labelled methionine incorporation in slices activated during G1-, S- and D-phases of the cell cycle of *H. tuberosus*. In the culture medium 15 µCi [3,4-^{14}C] methionine were added for the indicated periods. Proteins were determined by the Lowry method [25].

	3 h	18 h	27 h
nmol[^{14}C]met•mg prot $^{-1}$	0.28	2.18	2.87

within 3 h, consistent with high turnover (see above); increasing leucine incorporation between 3 and 6 h also occurs (Table 1 and Fig. 8). This protein synthesis was inhibited completely by the presence of 100 µM cycloheximide in the growth medium (Fig. 9). After mid-G1-phase, protein synthesis prevailed and thus protein levels increased rapidly (Fig. 5). The *in vivo* incorporation of [^{14}C] PU administered in the growth medium is demonstrated by the recovery of radioactivity in electrophoretic bands in the high molecular mass zone and in bands of 17.5 and 19 kDa [39]. Total TGase activity assayed in the cell-free

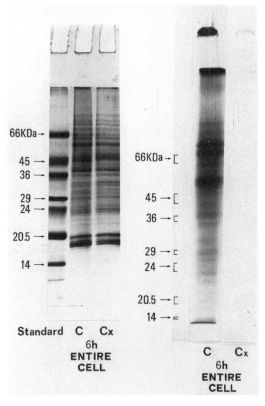

Fig. 9. Coomassie-stainable protein bands analyzed by SDS-PAGE [22] from cells of explants of *H. tuberosus* cultured for 6 h as reported in Fig. 3, in the presence or absence of 100 µM cycloheximide (left). The relative fluorographs of proteins of the same explants cultured in medium supplemented with 15 µCi [^{14}C] leucine per 5 g of fresh tissue are shown on the right-hand side.

extract decreased rapidly from the beginning of G1-phase until mid-G1 (Fig. 5); Ts TGase activity remained low [38]. Another early metabolic event is DNA synthesis in mitochondria and plastids which peaks around 3–6 h [36, 44]. Thus, organelle and nuclear DNA replication are independent. All three PA are also present in mitochondria [44] as well as their precursors and their biosynthetic enzymes [46].

After mid-G1 (the 6th hour), all the degradative processes observed in the first half of G1-phase seem to end. By contrast, the biosynthesis of free PA continues and reaches a maximum at the end of G1-phase (Fig. 4) [36]. ODC, ADC and DAO activities all increase considerably (Fig. 3). TGase activity increased considerably due to the synthesis of both Ts TGase and its substrates [16]. The products of TGase activity were analyzed on two dimensional TLC plates. Apart from PU, which decreased until mid-G1 and then increased in the next phases, bound PU derivatives, especially compounds having a migration near zero, were also formed. Very low amounts of labelled bound SD, SM or γ-aminobutyric acid were detected (Figs. 2) [9].

5. S-phase

When the synthesis of nuclear DNA begins, a decrease in PA content takes place (Figs. 3, 4); however the synthesis of PU, SD and SM still continues as shown by the *in vivo* incorporation of arginine [36], and by the increasing activities of biosynthetic enzymes. After mid-S-phase, all PA start to accumulate again up to the end of this phase (Figs. 2, 4) [36]. The relationship between PA and DNA synthesis has been studied in animal systems, where ODC, the only enzyme responsible for PU biosynthesis, can be inhibited by various factors (e.g. colchicine, vinblastin) that delay or block the S-phase. Moreover many inhibitors of PA biosynthesis block the cell cycle and their effect seems to be specific [17]. However, the effects of polyamine biosynthesis inhibitors on DNA are difficult to understand because frequently, especially in plants, only a partial depletion of PA is reached. Also the block of the cell cycle occurs gradually only after some doubling time periods as, for example, in *Chlorella vulgaris* [8]. In *H. tuberosus*, inhibitors of PA synthesis not only reduce the concentration of PA, but also affect DNA replication. However, the effect of these inhibitors on PA biosynthesis *in vivo* is transient and not very marked. Thus, there is little information on the consequences of PA depletion on the cell cycle [34]. Another approach was used to study the effect of spermidine (SD) on the DNA of activated slices of *H. tuberosus*. *In vivo* SD was able to partially alleviate the inhibition of actinomycin D on DNA and RNA synthesis. The direct interaction between DNA, actinomycin D and SD was confirmed by *in vitro* studies on their binding affinity [9]. Different models of the binding of PA to DNA by hydrogen linkages were proposed [13, 23].

The protein content and pattern of conjugates as well as TGase activity, previously discussed, followed the same trend observed after mid-G1-phase and reached the highest levels during mid-S-phase (Figs. 5, 8 and Table 1) [28, 39]. In S-phase the contents of high molecular mass conjugates obtained by *in vivo* incorporation of labelled PU increased [39], similarly to those obtained by incubating the cell-free extracts (Fig. 5).

6. Cell division

This phase largely overlaps with the S-phase (Fig. 3); consequently, some changes in PA metabolism can be related to events linked to either or both phases. The second peak in PA accumulation and in PU synthesis from arginine, together with high SD synthesis, occurs in the middle of the division phase (Fig. 4) [36]. During the last part of this phase it seems that PA biosynthesis is no longer necessary. The sharp decline in PA content might also result from a degradative metabolism due to the resumption of DAO activity (Figs 2, 3, 4).

   Protein levels begin to decrease (Fig. 5) and the rate of [^{14}C]-methionine incorporation was also lower than before (Table 1). The electrophoretic pattern was quite different with respect to that observed in G0-phase; in particular the percentage of high molecular mass protein bands increased considerably whereas other protein bands disappeared completely (Figs. 6, 7). All of these modifications were blocked by cycloheximide [16]. The formation of these large complexes is accompanied by a heavier PU incorporation both *in vivo* and in cell-free extracts, suggesting that TGase activity might play a role in the formation of these conjugates. Some of them can be separated on SDS-PAGE either by modulating the percentage of acrylamide from 12% to 10% or by digesting them with papain or a mixture of cellulase and pectinase [9]. In this phase, fibrillar material from the cytoplasm migrated into the walls of *H. tuberosus* cells [12] and in *Catharanthus roseus,* wall protein synthesis has been reported [20]. In animals, similar heavy aggregates due to TGase activity, were found [4]; common substrates for this enzyme appear to be extracellular structural proteins or intracellular structural proteins, like actin and myosin [24]. A functional dependence on PA for the formation of cytoskeletal structures, like actin filaments and microtubules, has been shown in PA-deficient CHO cells [30]; in addition, the PA depletion caused a high incidence of binucleate fibroblasts [43]. In plants, PA may also be involved in the formation of supramolecular cell structures and of cell walls, whose formation is one of the main events of cytokinesis. PA have been found located in the cell walls [27, 32], where they form ionic linkages with pectic substances [10] and hydrogen bonds with neutral wall polysaccharides.

7. Conclusion

To date, most studies have dealt only with free PA, but their exact role in the cell has not been clarified. The similarity between the bimodal trend of free PA content in *H. tuberosus* [36], *N. tabacum* [29] and *Catharanthus roseus* [26] compared with that found in animal cells [18] is clear. The behaviour of SAMDC (two maxima of activity in G1- and S-phases) and of ODC (a peak in early S) are very similar in *H. tuberosus, C. roseus* and in animal tissues [17, 18, 26] which suggests that PA follow a common program and play an important role in cell biology. The main difference is the occasional presence of a maximum of ODC activity in G1 observed only in some animal cells (Fig. 4) [17].

It is sometimes difficult to interpret endogenous PA levels since they can be stored (in vacuoles or cell walls) where they cannot interact with the metabolic pathways occurring in the cytoplasm or nucleus. While free PA follow a similar trend in *H. tuberosus* and *N. tabacum*, the amount of TCA-insoluble PA detected in *N. tabacum* follows a bimodal trend and that of the conjugates in *H. tuberosus* shows a sigmoidal trend. This again confirms that these conjugates cannot be compared with TCA-insoluble bound PA as a whole, since in the latter many different molecules could bind to PA. In the conjugates, the partner molecules are proteins, even though, to date, they have not been clearly identified.

The concentration of labelled PU conjugated *in vivo* to high molecular mass proteins during the cell cycle, is $0.2 \ \mu M$. Thus, even taking into account the dilution by internal PU (from 1 to 60 μM according to the phases), the percentage of conjugates is not relevant; perhaps their role in post-translational protein modification is critical. It would be worthwhile to identify the role of each different PA. Clearly, PA in the cell should not be considered simply as a single entity but as distinct categories: free, non-tightly-bound, tightly-bound or covalently linked.

The experiments performed with *H. tuberosus* suggest some possible roles of PA during the cycle: 1) PA are necessary for DNA duplication as well as for RNA and protein synthesis and enhance the efficiency and fidelity of these processes; 2)PA can be involved in the "activation" of RNA and DNA or in protecting them from nucleases; 3) PA may be necessary for the cross-linking of cytoskeletal or wall proteins (or, possibly, of enzymatic ones) via covalent linkages; 4) PA form hydrogen and ionic linkages with cell wall polysaccharides and are thus important for cell wall properties; 5) PU may be necessary, as a precursor of GABA, to furnish organic acids for the Krebs cycle and amino acids for protein synthesis.

Acknowledgements

We thank both past and present members of the lab, and Dr. Stefania Biondi for reviewing the manuscript.

The research was supported by National Research Council of Italy, Special Project RAISA, Sub-project N.2, paper N...

154

References

1. Bagni N (1966) Aliphatic amines as a growth-factor of coconut milk as stimulating cellular proliferation of *Helianthus tuberosus* (Jerusalem artichoke) *in vitro*. Experientia 22: 732–736.
2. Bagni N, Stabellini G and Serafini-Fracassini D (1973) Polyamines bound to tRNA and rRNA of eukaryotic plant organisms. Physiol Plant 29: 218–222.
3. Bagni N and Torrigiani P (1991) Polyamines: a new class of growth substances. In: Proc. of the 14th Conference on Plant growth substances. pp. 264-275 Dordrecht (The Netherlands): Kluwer Academic Publishers.
4. Beninati S, Piacentini M, Cocuzzi ET, Autuori F and Folk JE (1988) Covalent incorporation of polyamines as γ-glutamyl derivatives in CHO cell protein. Biochim Biophys Acta 952: 325–333.
5. Bertossi F, Bagni N, Moruzzi G and Caldarera CM (1965) Spermine as a new promoting substance for *Helianthus tuberosus* (Jerusalem artichoke) *in vitro*. Experientia 21: 80–82.
6. Bonner J and Addicott F (1937) Cultivation *in vitro* of excised pea roots. Bot Gaz 99: 144–170.
7. Bradford MM (1976) A rapid and sensitive method for the quantitation of microgram quantities of protein utilizing the principles of protein-dye binding. Anal Biochem 78: 248–254.
8. Cohen E, (Malis) Arad S, Heimer YH and Mizrahi Y (1984) Polyamine biosynthetic enzymes in the cell cycle of *Chlorella*. Correlation between ornithine decarboxylase and DNA synthesis at different light intensities. Plant Physiol 74: 385–388.
9. Dinnella C, Serafini-Fracassini D, Grandi B and Del Duca S (1992) The cell cycle in *Helianthus tuberosus*. Analysis of polyamine-endogenous protein conjugates by TGase-like activity. Plant Physiol Biochem 30 (5): 531-539.
10. D'Orazi D and Bagni N (1987) *In vitro* interactions between polyamines and pectic substances. Biochem Biophys Res Commun 148: 1259–1263.
11. D'Orazi D, Serafini-Fracassini D and Bagni N (1979) Polyamine effects on the stability of DNA-actinomycin D complex. Biochem Biophys Res Commun 90: 362–367.
12. Favali A, Serafini-Fracassini D and Sartorato P (1984) Ultrastructure and autoradiography of dormant and activated parenchyma of *Helianthus tuberosus*. Protoplasma 123: 192–202.
13. Feuerstein BG, Basu HS and Marton LJ (1988) Theoretical and experimental characterization of polyamine/DNA interactions. In: Progress in Polyamine Research. V Zappia and AE Pegg (eds) Novel Biochemical, Pharmacological, and Clinical Aspects. Vol 250 of Advances of Experimental Medicine and Biology, pp. 517–523. New York and London: Plenum Press.
14. Folk JE (1980) Transglutaminases. Annu Rev Biochem 49: 517–531.
15. Galston AW and Kaur-Sawhney R (1990) Polyamines in plant physiology. Plant Physiol 94: 406–410.
16. Grandi B, Del Duca S, Serafini-Fracassini D and Dinnella C (1992) Re-entry in cell cycle: protein metabolism and transglutaminase-like activity in *Helianthus tuberosus*. Plant Physiol Biochem 30 (3): 415-424.
17. Heby O (1981) Role of polyamines in the control of cell proliferation and differentiation. Differentiation 19: 1–20.
18. Heby O, Gray JW, Lindl PA, Marton LJ and Wilson CB (1976) Changes in L-ornithine decarboxylase activity during the cell cycle. Biochem Biophys Res Commun 71: 99–105.
19. Icekson I and Apelbaum A (1987) Evidences for transglutaminase activity in plant tissue. Plant Physiol 84: 972–974.
20. Kodama H, Ito M and Komamine A (1990) Molecular aspects of the cell cycle in *Catharanthus roseus* synchronous cell division cultures. In: HJJ Nijkamp, LHW Van der Plas and J Van Aartrijk (eds). Progress in Plant Cellular and Molecular Biology, pp. 532–536. Dordrecht: Kluwer Academic Publishers.
21. Korner G and Bachrach U (1987) Intracellular distribution of active and inactive transglutaminase in stimulated cultured C6 glioma cells. J Cell Physiol 130: 44–50.
22. Laemmli UK (1970) Cleavage of structural proteins during the assembly of the head of bacteriophage T4, Nature 227: 680–683.

23. Liquori AM, Costantino L, Crescenzi V, Elia V, Giglio E, Puliti R, De Santis-Savino M and Vitagliano V (1967) Complexes between DNA and polyamines: A molecular model. J Mol Biol 24: 113–122.

24. Lorand L and Conrad SM (1984) Transglutaminases. Mol Cell Biochem 58: 9–35.

25. Lowry OH, Rosebrough NJ, Farr AL and Randall RJ (1951) Protein measurement with the Folin phenol reagent. J Biol Chem 193: 265–275.

26. Maki H, Ando S, Kodama H and Komamine A (1991) Polyamines and the cell cycle of *Chatharanthus roseus* cells in culture. Plant Physiol 96: 1008–1013.

27. Mariani P, D'Orazi D and Bagni N (1989) Polyamines in primary walls of carrot cells: endogenous content and interactions. J Plant Physiol 135: 508–510.

28. Mossetti U, Serafini-Fracassini D and Del Duca S (1986) Conjugated polyamines during dormancy and activation of tuber of Jerusalem artichoke. In: Conjugated Plant Hormones. K Schreiber, HR Schütte and G Sembdner (eds) Structure, Metabolism and Function, pp. 369–375. Berlin: Deutscher Verlag der Wissenschaften.

29. Pfosser M, Königshofer H and Kandeler R (1990) Free, conjugated, and bound polyamines during the cell cycle of synchronized cell suspension cultures of *Nicotiana tabacum*. J Plant Physiol 136: 574–579.

30. Pohjanpelto P, Virtanen I and Hölttä (1981) Polyamine starvation causes disappearance of actin filaments and microtubules in polyamine-auxotrophic CHO cells. Nature 293: 475–477.

31. Quigley GJ, Teeter MM and Rich A (1978) Structural analysis of spermine and magnesium ion binding to yeast phenylalanine transfer RNA. Proc Natl Acad Sci USA 75: 64–68.

32. Scoccianti V, Bagni N, Dubinsky O and Arad (Malis) S (1989) Interaction between polyamines and cells of marine unicellular red alga *Porphyridium* sp. Plant Physiol Biochem 27: 899–903.

33. Seiler N (1983) Liquid chromatographic methods for assaying polyamines using prechromatographic derivatization. Methods Enzymol vol. 94, Polyamines, 10–25.

34. Serafini-Fracassini D., 1991. Cell Cycle-Dependent Changes in Polyamine metabolism. In: Biochemistry and Physiology of Polyamines in Plants. RD Slocum and HE Flores (eds), pp. 175–186. CRC Press Uniscience, Boca Raton, Ann Arbor, London.

35. Serafini-Fracassini D and Alessandri M (1983) Polyamines and morphogenesis in *Helianthus tuberosus* explants. In: Advances in Polyamine Research, Vol. 4. U Bachrach, AM Kaye and P Chayen (eds) pp. 419–426. New York: Raven Press.

36. Serafini-Fracassini D, Bagni N, Cionini PG and Bennici A (1980) Polyamines and nucleic acids during the first cell cycle of *Helianthus tuberosus* tissue after the dormancy break. Planta 148: 332–337.

37. Serafini-Fracassini D, Del Duca S and D'Orazi D (1988a) First evidence for polyamine conjugation mediated by an enzymic activity in plants. Plant Physiol 87: 757–761.

38. Serafini-Fracassini D, Del Duca S and Torrigiani P (1989) Polyamine conjugation during the cell cycle of *Helianthus tuberosus*: non enzymatic and transglutaminase-like binding activity. Plant Physiol Biochem 27: 659–668.

39. Serafini-Fracassini D, Del Duca S, D'Orazi D and Mossetti U (1988b) Conjugation of polyamines by an enzyme activity in dividing cells of *Helianthus tuberosus*. In: Perspectives in Polyamine Research. A Perin, D Scalabrino, A Sessa and L Ferioli (eds) pp. 89–92. Milano: Wichtig Editore.

40. Serafini-Fracassini D, Torrigiani P and Branca C (1984) Polyamines bound to nucleic acids during dormancy and activation of tuber cells of *Helianthus tuberosus*. Physiol Plant 60: 351–357.

41. Signorini M, Beninati S and Bergamini C (1991) Identification of transglutaminase activity in the leaves of Sugar Beet (*Beta vulgaris* L.). J Plant Physiol 137: 547–552.

42. Smith TA, Negrel J and Bird CR (1983) The cynnamic acids amides of the di- and polyamines. In: U Bachrach, AM Kaye and P Chayen (eds) Advances in Polyamine Research, vol. 4 pp. 347–370. New York: Raven Press.

43. Sunkara PS, Rao PN, Nishioka K and Brinkley BR (1979) Role of polyamines in cytokinesis of mammalian cells. Exp Cell Res 119: 63–68.

44. Torrigiani P and Serafini-Fracassini D (1980) Early DNA synthesis and polyamines in mitochondria from activated parenchyma of *Helianthus tuberosus*. Z Pflanzenphysiol 97: 353–359.
45. Torrigiani P, Altamura MM, Capitani F, Serafini – Fracassini D. and Bagni N (1989) *De novo* root formation in thin cell layers of tobacco: changes in free and bound polyamines. Physiol Plant 77: 294–301.
46. Torrigiani P, Serafini-Fracassini D, Biondi S and Bagni N (1986) Evidence for the subcellular localization of polyamines and their biosynthetic enzymes in plant cells. J Plant Physiol 124: 23–29

12. Cytoskeletal analysis of maize meiotic mutants

CHRISTOPHER J. STAIGER and W. ZACHEUS CANDE

Abstract

Microsporogenesis is an ideal system for studying cell division and the role of the cytoskeleton during higher plant development. Male meiosis in maize has been studied extensively using conventional cytological techniques, and more recently by fluorescence microscopy. The collection of 26 existing meiotic mutants can be grouped into six categories including: meiotic initiation, prophase chromosome pairing, chromosome morphology, chromosome segregation and cytokinesis, variable meiotic defects, and exit from meiosis. Using indirect immunofluorescence staining we have characterized the timing and nature of perturbations to the microtubule cycle in nine meiotic mutants. Several have microtubule disruptions that precede any other cytological abnormalities. For example, progression from a prophase microtubule array to a metaphase spindle is abnormal in the mutant $dv1$. Instead of converging to form focused poles, the metaphase spindle poles remain broad and divergent as in prometaphase. Two other mutants, $ms17$ and $ms28$, have defects including abnormal or multiple spindles, and cytokinetic failure that can be attributed to an excess of microtubules . The recessive mutation $ameiotic1$ replaces meiosis I with a mitotic division. All features of these sporocytes, including chromosome behaviour, are typical of a somatic division. Surprisingly, a preprophase band predicts the future plane of division. These observations suggest that the wild type function of $Ameiotic$ is to superimpose unique features of meiosis onto a mitotic division. One advantage of the microsporogenesis system is the availability of powerful genetic tools including transposon tagging. We have recovered several recessive meiotic mutants from Robertson's *Mutator* lines. One mutation ($dsy*$) is similar to existing synaptic mutants and the other ($aph1$) has a novel phenotype with interesting effects on phragmoplast formation and positioning.

1. Introduction

Cell division is a complex process that is fundamental to the survival and propagation of all organisms, including higher plants. While it can be assumed that karyokinesis and cytokinesis require the action of a large number of genes and gene products, very few cytological components involved in these processes

157

J.C. Ormrod and D. Francis (eds.), Molecular and Cell Biology of the Plant Cell Cycle, 157–171.
© 1993 *Kluwer Academic Publishers. Printed in the Netherlands.*

have been identified [9, 36]. Microtubules and microfilaments, as major components of the spindle and phragmoplast, are certain to play an important role during cell division [33, 50], yet how these structural elements function in chromosome separation and cytokinesis awaits the discovery of essential interacting proteins. The genetic control of cell division has been elegantly studied in many non-plant systems including yeast and other fungi, *Drosophila*, and mammalian cells (reviewed in [2, 9, 36]). Mutational analysis of these organisms led to the discovery of regulatory cascades, and structural components necessary for cell division (see J Hayles and P Nurse, this volume). This approach is informative because it lets the organism (or cell) tell us what genes or gene products are important. A similar strategy should be useful for identifying both structural and regulatory components in higher plant cell division.

Here we describe the development of a system to study higher plant cell division which has the advantage of combining powerful genetic analysis with current cytological techniques. Because null mutations in genes controlling mitotic cell division are likely to be lethal, one alternative is to examine genetic disruptions of meiosis. Although mitosis and meiosis differ in signicant ways, especially regarding prophase chromosome behaviour and kinetochore function, they are strikingly similar in other respects. For example, chromosomes are segregated on a bipolar spindle composed primarily of microtubules and possibly also actin, and cell plate formation is mediated by a typical phragmoplast [51, 52]. Microsporogenesis is an unexploited system perfectly suited for studying the genetic regulation of microtubule organization, chromosome segregation and cytokinesis. Maize is a classic genetic organism with 1) a superb cytogenetic characterization of meiosis, 2) a large collection of mutants, many of which are disturbed in chromosome segregation or cytokinesis, 3) the ability to use transposon mutagenesis to generate new mutations and to subsequently clone them by transposon tagging. In this paper, a survey of the large collection of existing maize meiotic mutants is presented. The cytoskeletal phenotype of several interesting meiotic mutations is compared to wild type microtubule patterns. A strategy for identifying genes and gene products important to meiosis is also described.

1.1. Maize microsporogenesis

Microsporogenesis is the process by which a diploid pollen mother cell, or primary microsporocyte, undergoes a meiotic reduction division followed by cytokinesis to yield four haploid microspores. Each microspore undergoes further development to produce the gametophyte generation, or mature pollen grain. This process occurs within the anthers of male florets on a large terminal inflorescence or tassel. Meiosis in maize has been described beautifully by many workers [10, 12, 32, 42, 43]. These descriptions usually focus on the complex behaviour of chromosomes, especially homologous chromosome pairing, synapsis, and chiasma formation, as well as chromosome segregation during the two meiotic divisions. Several attributes make this process an excellent choice for the study of cell division. Maize anthers are quite large (3–5 mm) and each contains about 100 synchronous sporogenous cells. Moreover, the development of anthers in a floret and along the tassel is quite predictable, simplifying the location of meiotic stages. The ability to mechanically extrude meiocytes from cut anthers facilitates rapid fixation and handling of meiocytes for immunofluorescence. Maize meiocytes are often greater than 50 μm in diameter. They lack a single large vacuole, and their callose wall is easily made permeable to probes such as antibodies. Furthermore, previous cytological work allows precise correlation of chromosome morphology with meiotic stage. As described elsewhere [51, 52, 53], these cells represent ideal material for examining cytoskeletal arrays during higher plant division.

1.2. Catalogue of maize meiotic mutants

A primary attraction of maize microsporogenesis is the large collection of mutants affecting all stages of development (see also [1, 5, 10, 14, 17, 19, 31]). Meiotic mutants are identified initially by their effect on pollen viability. Mutations that cause pollen sterility are the most common type of naturally occurring and induced genetic lesions in flowering plants (reviewed in [31]). A survey of the literature through 1991 shows 42 maize mutations affect pollen fertility, 29 of which have been placed on a specific chromosome. Approximately a third [18] of the sterility mutations disrupt development after the microspore stage, and can be considered genes supporting male gametophyte formation. The rest (26 mutations) are defective at various stages of meiosis (Table 1). The largest class of meiotic mutants (seven in total) affects chromosome pairing and synapsis during prophase I. Six mutations affect meiotic chromosome segregation or cytokinesis. Many from this last class were selected for cytoskeletal examination because they disturb basic cell division processes.

All maize meiotic mutations are inherited as recessive characters, except for the dominant trait *Mei*025. Although the collection of meiotic mutations is quite large, there are only multiple alleles for two genes (loci), *ameiotic* (*am*1 = *am-pra*1) (I. Golubovskaya, personal communication) and *polymitotic* (*po* = *ms*4 = *ms*6). This indicates that we have not yet identified all the genes required for normal meiosis and pollen development. However, not all mutant genes have been tested for the ability to complement. While tests for allelism between 16 male sterile loci were conducted by Beadle [5], many mutations remain unmapped, and mutations on the same chromosome but with quite different cytological descriptions may not have been crossed. Seed stocks for most of these mutations can be obtained through the Maize Stock Center (E Patterson, University of Illinois). In addition, a little known collection of meiotic mutants has generously been made available by Dr. Inna Golubovskaya (N.I. Vavilov Institute of Plant Industry, St. Petersburg, Russia).

Table 1. Meiotic genes of maize

Mutant	Location	Phenotype	References
MEIOTIC INITIATION			
am	5S-20	*ameiotic*; meiosis replaced by a mitotic	
*(am-pra*1)		division	[21, 22, 40, 54]
CHROMOSOME PAIRING AND SYNAPSIS			
afd	6L	*absence of first division*; failure of synapsis,	
		univalents present at MI, some SC formation	[18, 22, 23, 26, 28, 34]
as	1S-56	*asynaptic*; failure of chromosome synapsis	[3, 6, 8, 22, 28, 37]
dy	nd	*desynaptic*; partial failure of chromosome	
		pairing, several univalents present at MI	[39]
*dsy*1	not 1 or 6L	*desynaptic*; premature terminalization of	
		chiasmas, univalents present at MI	[19, 22, 24, 26, 28, 34]
*dsy*2	nd	*desynaptic*; like *dsy*1	[19, 28, 34]
*dsy*4	nd	*desynaptic*; like *dsy*1	[19]
dsy*	nd	*desynaptic*; like *dsy*1	this study
CHROMOSOME STRUCTURE AND MORPHOLOGY			
st	4S-62	*sticky chromosome*; anaphase chromatin	
		bridges	[7, 47]
*Mei*025	nd	hyper-condensation of chromatin at PI-MI,	
		failure to segregate chromosomes in AI	[27, 34]
el	8L	*elongate*; premature despiralization of	
		chromosomes at both meiotic divisions	[44]
*K*10	10L-near *sr*2	*abnormal*10, induces neocentromere activity	
		on knobs	[45]
SPINDLE OR CYTOKINESIS			
dv	not 1S, 5,	*divergent spindle poles* (not allelic to *ms*28,	
	6S, 7L, 9S	*ms*43, or *ms*17)	[13, 48, 51, 52]
*ms*17	1S-23	abnormal spindle formation, variable	
		expression	[1, 15, 52]
*ms*28	1S	hyperstable spindle, poor chromosome	
		segregation, partial or total cytokinesis	
		failure	[18, 27, 34, 48, 49]
*ms*43	8L	wide and weak spindle, abnormal	
		chromosome segregation, chromosomes fail	
		to decondense	[20, 27, 34, 48, 49]
va	7L-near *ij*	*variable sterile*, partial or total cytokinesis	
		failure after meiosis I	[4, 5]
*aph*1	nd	*abnormal phragmoplast* formation	this study
VARIABLE STAGE OF EXPRESSION BEFORE OR DURING MEIOSIS			
*pam*1	6L ??	*plural abnormalities of meiosis*; various	
		defects at stages from premeiosis through	
		pollen development	[19, 25, 28, 34]
*pam*2	nd	*plural abnormalities of meiosis*; like *pam*1	[19, 28]
*ms*22	nd	PMC breakdown during early PI (not allelic	
		to *ms*8,9 or 17)	[55]
*ms*8	8L-28	variable PMC breakdown, often as early as	
		PI	1, 5, 16

Continued from *Table 1.*

Mutant	Location	Phenotype	References
*ms*9	1S-near *P*	partial failure of cytokinesis, cytoplasmic breakdown like *ms*8	[1, 5, 29]
*ms*23	3L	allelic to *ms*-Bear*7, variable cytoplasmic breakdown, as early as PI, like *ms*8	[5]
EXIT FROM MEIOSIS			
po	6S-4	*polymitotic*, supernumerary mitoses after	
*(ms*4,*ms*6)		meiosis II	[1, 28]

2. Materials and methods

Seeds from maize inbred strains B73, and A344, as well as from mutant stocks *dv*, *el*, *am*1, *ms*8, *ms*9 (Maize Genetics Stock Center, University of Illinois) and *Mei*025, *ms*28, *ms*43 (Dr. Inna Golubovskaya, N.I. Vavilov Institute of Plant Industry) and *ms*17 (gift of Dr. Marc Albertson, Pioneer Hi-Bred International) were germinated in vermiculite flats for 2–3 weeks before transfer to soil and grown to maturity in the greenhouse. The staging of plant material, tassel removal, and manipulation of microsporocytes for indirect immunofluorescence microscopy have been described previously [51, 52, 53].

3. Results and discussion

3.1. Microtubule organization during wild type microsporogenesis

We have analyzed the spatial and temporal changes in microtubule distribution during wild type maize microsporogenesis using indirect immunofluorescence microscopy [51, 53]. Interphase, prophase and telophase cytoplasmic microtubule arrays emanate from the nuclear envelope and radiate towards the cortex, supporting the theory that a site closely associated with the nuclear envelope serves as an MTOC (microtubule organizing centre). Meiocytes lack both an organized cortical array of microtubules such as that found in somatic interphase cells, and the preprophase band (PPB) of microtubules that marks the future division plane of somatic cells (also see D Francis and RJ Herbert, this volume). However, during meiosis there is rigorous control over division polarity. In maize and other grasses, the orientation of the meiotic division plane is very predictable and apparently determined before chromosome separation (reviewed in [11, 42, 51]). Meiotic spindles during both divisions have highly focused poles and a regular orientation within the cells and the anther locule. Cytokinesis immediately follows both meiotic karyokineses and is effected by a typical phragmoplast during late anaphase and telophase. The parallel arrays of phragmoplast microtubules propagate centrifugally forming a ring around the newly deposited cell plate, and cytokinesis is always completed before the next division ensues. The first division results in a dyad of secondary microsporocytes. An isobilateral tetrad of coplanar microspores is the ultimate product of the second meiotic division. Only in rare instances (<1% of cells) are meiotic anomalies observed in wild type material.

3.2. Cytological and cytoskeletal defects in meiotic mutants

Genes controlling the progression through meiosis can be grouped into several classes according to their time and mode of action (see Table 1). These categories include; initiation of meiosis (one gene, two mutant alleles), prophase chromosome pairing and synapsis (seven mutants), chromosome morphology (four mutants) meiotic division including chromosome segregation and cytokinesis (six mutants), variable stages of expression (six mutants), and exit from mitosis (one gene, three alleles). Nine of these existing meiotic mutants were examined during the course of this study (*am*1, *Mei*025, *el*, *dv*1, *ms*17, *ms*28, *ms*43, *ms*8, and *ms*9). Additionally, we have isolated and characterized two new meiotic mutations. The first mutation isolated, *dsy**, causes desynapsis and the second, *aph*1, leads to an abnormal proliferation of cytokinetic structures or phragmoplasts (see later). Cytoskeletal analysis reveals that many of the existing meiotic mutants have defects which precede any previously reported defects (*am*1, *dv*1, *ms*17, *ms*28). A brief synopsis of cytological and cytoskeletal abnormalities associated with the various mutant classes follows.

Ameiotic (*am*1, *am-pra*1) acts earlier than any maize sterility gene leading others to propose that the wild type gene regulates the initiation of meiosis [19]. Homozygous *am*1/*am*1 sporocytes undergo a synchronous division but this ameiotic division is predominantly mitotic in nature [40]. For example, sporocytes show no chromosome pairing, synaptonemal complex formation, or any other chromosome behaviour typical of meiotic prophase I [19]. Furthermore, 20 univalents (not 10 bivalents) are aligned on the metaphase plate and sister chromatids separate at anaphase I [40]. The *ameiotic1* mutation does not alter the timing of cell division. The ameiotic division occurs concurrent with the ultimate tapetal division which normally takes place when sporocytes are in prophase I [54]. Cytoskeletal examination of *am*1/*am*1 sporocytes revealed that microtubule organization resembled mitosis rather than meiosis, including the presence a PPB (Fig. 1A), [54]. The presence of a PPB in *ameiotic* sporogenous cells suggests that its elimination from normal meiosis is genetically regulated. These observations and published genetic studies [19] lead us to propose a model whereby the wild type *Ameiotic* gene acts during G2 or earlier and is necessary for maintaining several independent meiotic processes, including both chromosome behaviour and microtubule organization. In simple terms, the function of *Ameiotic* is to superimpose meiotic features onto a mitotic cell cycle [54]. This implies that a PPB-

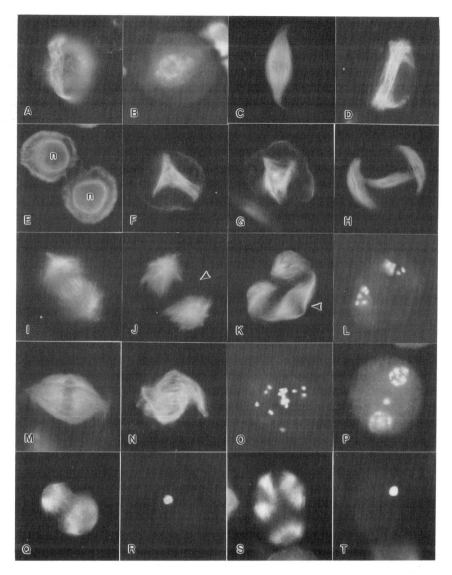

Fig. 1. Microtubule distribution and chromosome behaviour in the maize meiotic mutants *am*1, *dv*1, *ms*17, *ms*28, *ms*43 and *Mei*025.

(A) A PPB in an *am*1/*am*1 meiocyte. (B) DAPI-stained image of the cell in A. (C) Wild type metaphase I spindle with highly focused poles. (D) *dv*1/*dv*1 spindle with divergent poles. (E) *ms*17/*ms*17 prophase I meiocyte with an extra cortical array and normal perinuclear (n) array. (F) Prometaphase I *ms*17/*ms*17 cell with multipolar spindle and cortical array. (G) Another prometaphase *ms*17 cell. (H) Multiple metaphase I spindles. (I) An ms28/ms28 spindle with over-abundant polar microtubules. (J) A telophase I *ms*28/*ms*28 cell lacking a phragmoplast (arrow) but containing pronounced radial arrays. (K) An *ms*28 cell with two metaphase II spindles and a phragmoplast from division I (arrow) in a common cytoplasm. (L) DAPI-stained image of the cell in K confirms that the phragmoplast has no associated chromosomes. (M) An extra-wide

---

metaphase I spindle in a *ms*43/*ms*43 sporocyte. (N) Premature spindle degeneration in *ms*43. (O) Abnormal chromosome segregation in the cell shown in N. (P) Chromosome decondensation fails in products of the *ms*43 division. (Q) A *Mei*025 cell with two radial interphase microtubule arrays but only one nucleus (R) positioned between them. (S) Three interphase arrays in another *Mei*025 sporocyte are not associated with the single, centrally-placed nucleus shown in (T). ($\times$ 700)

containing mitotic cell cycle is the default condition for plant cell division.

Homologous chromosome pairing and recombination are key events in meiotic prophase controlled by at least seven different genes (Table 1). Because each of these mutations also affects female fertility we predict that these processes are regulated by a single set of genes functional in both female and male meiosis. Chromosome pairing has been shown to correlate with the presence of a specific structure, the synaptonemal complex (SC), visible in ultrastructural studies of prophase chromosomes [38]. All of the prophase mutants form an SC, but it is either not extensive (*afd*, *as*) or its destruction is premature (*dsy*1) (reviewed in [19]). Abnormalities of the SC are correlated with subsequent defects in chromosome behaviour, and all prophase I mutants have univalents present at diakinesis. For example, univalents can cause abnormal spindle formation (e.g. *as*, *dsy*1, and *dsy*** see later) and often fail to segregate properly in anaphase I [24, 8, 3, 37]. In addition, centromere separation may occur prematurely resulting in univalent chromatids that segregate equationally during division I rather than division II (e.g. *afd*, *dy*, *as*; references 23, 37, 19).

A third class of meiotic mutations alters chromosome morphology or structure (Table 1). Unusually sticky chromosomes can be observed as early as pachytene in the dominant mutant *Mei*025 ,or as late as anaphase in *sticky* (*st*) [7, 27, 34]. *Mei*025 sporocytes also have interesting cytoskeletal disruptions perhaps due to the inability to normally segregate the hyper-condensed chromatin mass, or alternatively due to pleiotropic effects of the mutation (Figs. 1Q-1T). Normal interphase and prophase microtubule arrays form as radial networks emanating from the nuclear envelope [51]. In *Mei*025 sporocytes typical nuclei do not form after meiosis I, but one or more radial arrays are always observed (Figs. 1Q, 1S). This suggests that microtubule organizing material can be dissociated from its normal location at the nuclear surface and assume other cytoplasmic postitions. Two other mutations, *K*10 and *el*, cause abnormal chromosome behaviour which only becomes visible during chromosome segregation. *Abnormal*10 (*K*10) induces neocentromeric activity on knobbed chromosomes leading to preferential chromosome segregation in the female [45]. *Elongate* (*el*) has abnormally stretched chromosomes aligned on the metaphase plate, and premature chromatin despiralization during anaphase [44]. Microtubule distribution in *el*/*el* microsporocytes is not affected by these changes in chromosome behaviour (data not shown).

Evidence for the genetic control of chromosome segregation and cytokinesis is provided by six mutations affecting these processes during maize microsporogenesis (Table 1). Three maize meiotic mutations disturb spindle formation and morphology (*dv*1, *ms*17, and *ms*28). These mutants have little or no effect on female fertility, consistent with the hypothesis that spindle formation, and chromosome segregation are regulated separately in male and female tissues. *Divergent spindle* (*dv*1), has abnormally broad spindle poles at metaphase I (compare Fig. 1D (*dv*1) with Fig. 1C (wt)), leading to abnormal anaphase chromosome segregation (see also [13, 51]). Because there is a disruption in the transition from the prophase microtubule array to a metaphase spindle, we believe this defect is caused by a perturbation of the microtubule organizing centre (MTOC) structure [51]. Spindle form is also altered in *ms*17/*ms*17 and *ms*28/*ms*28 homozygotes ([1, 52] CJ Staiger and WZ Cande, unpublished). Although these mutations are strikingly similar and both map to the short arm of chromosome 1 (Table 1), they are not alleles of the same gene (CJ Staiger and WZ Cande, unpublished). Both have extra, unexpected microtubule arrays during meiosis. Homozygous *ms*17 sporocytes have an unusual cortical microtubule array which first appears in late prophase (Fig. 1E) and often persists throughout meiosis (Figs. 1F, 1G). The prometaphase array is multipolar rather than bipolar (Figs. 1F,1G) and leads to multipolar and multiple spindles at metaphase I (Fig. 1H) The first microtubule abnormality in *ms*28 sporocytes occurs later coincident with poor chromosome segregation [27, 48]. Anaphase I spindles have excessive polar microtubules (Fig. 1I) giving them the appearance of an animal-type astral spindle. This observation confirms results from an earlier study using polarized light microscopy [49] and demonstrates that *ms*28 spindles contain more microtubules than wild type. Other microtubule abnormalities in *ms*28 sporocytes include failure to form normal phragmoplasts (Fig. 1J) and delay in their disassembly (Figs. 1K, 1L). We conclude that all of these defects stem from abnormal numbers of microtubules and/or a change in microtubule dynamics.

A recessive mutation in the *Ms*43 gene not only causes abnormalities in spindle structure but also modifies chromosome behaviour (see also [20, 27]). The first difficulties occur at MI when spindles appear abnormally wide (compare Fig. 1M (*ms*43) with Fig. 1C (wt)). Spindle structure is altered dramatically during anaphase giving the impression of premature disintegration (Fig. 1N). Coincident with spindle degeneration, and perhaps as a consequence of it, chromosome segregation is aberrant (Fig. 1O). Cytokinesis usually fails in these cells due to the absence of phragmoplast formation. Post-division defects in *ms*43 are not limited to cytoskeletal organization. Significantly, the meiotic chromosomes do not decondense from their anaphase-like state (Fig. 1P), although they do sometimes reform nuclei. Two possible models could explain the behaviour of *ms*43 sporocytes. First, the defects could be attributed to an altered spindle structural component required for the maintenance of spindle architecture and proper chromosome segregation. Alternatively, *ms*43 may be deficient for some regulatory component required for the normal progression through anaphase and re-entry into interphase. This latter model is consistent with the multiplicity of effects encompassing both cytoskeletal organization and meiotic chromosome behaviour.

Variable stages of expression are characteristic of six recessive meiotic mutations (Table 1). Two genes, termed *plural abnormalities of meiosis* (*pam1*, *pam2*), are the most general and often have premeiotic effects leading to coenocytic sporocytes [19, 25, 34]. Cytoplasmic breakdown during prophase I or later is caused by four different genes *ms8*, *ms9*, *ms22*, and *ms23* [1, 5, 16, 29, 55]. Nuclear behaviour is normal in each of these mutations. Cytoskeletal analysis of *ms8*, and *ms9* revealed no obvious defects in cytoplasmic, spindle, or phragmoplast microtubules during the first meiotic division even in cells where the cytoplasm was obviously degenerating (data not shown).

The final category of meiotic mutation is represented by multiple alleles for a single gene. Recessive mutations of *Polymitotic* (*po*) allow normal meiotic progression until the tetrad stage [1]. Rather than entering an extended interphase and depositing the gametophytic cell wall, *po/po* microspores undergo several successive cycles of chromosome segregation and cytokinesis without chromosome replication (QQ Liu and WZ Cande, unpublished; [1]). This suggests that *po* causes a defect in cell cycle regulation that blocks exit from meiosis and allows for the reiteration of meiosis II.

3.3. New meiotic mutants generated by transposon mutagenesis

The potential for identifying genes and gene products involved in higher plant cell division was the major reason for choosing maize microsporogenesis as a model system. Existing maize sterility mutations arose spontaneously or were induced by chemical mutagenesis; cloning by conventional methods will therefore be difficult. A powerful alternative strategy making use of transposon insertion can be used to create new mutations and eventually clone the gene of interest. Several different maize transposable element systems have been used to generate new developmental mutants and to clone previously characterized genes [41, 35, 30, 46].

For this study we used the Robertson's *Mutator* transposable element system to generate mutations. Of 2000 F_2 families screened, 64 segregated for male sterility. These were replanted in a subsequent field season, and sporogenous material examined for meiotic defects. Based on this cytological screen 17 F_2 families were selected for further study. To establish the genetic nature of these meiotic lesions, putative homozygous sterile plants were test-crossed to an inbred and the progeny selfed to recover the mutation. Families of 20 plants each were screened cytologically for the segregation of meiotic defects. Heritability studies indicated that five families segregated meiotic lesions as recessive traits. Our success demonstrates the feasibility of this approach for identifying novel meiotic mutations; however, it is a very laborious and lengthy process. A brief description of two new meiotic mutations follows.

The mutation *dsy** is a monogenic recessive trait that causes chromosome

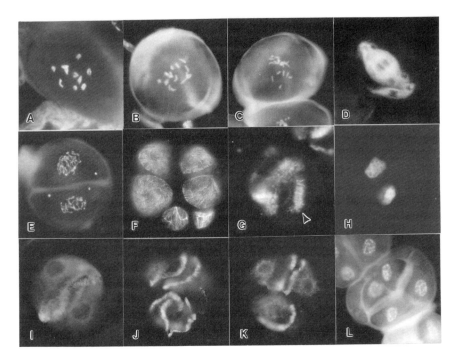

Fig. 2. Two new meiotic mutants generated by transposon mutagenesis; *dsy** and *aph*1.
(A) Wild type meiocyte during diakinesis shows 10 homologous chromosome pairs. (B&C) *dsy*/dsy** cells with 2 or more univalents or unpaired chromosomes at diakinesis. (D) Metaphase I *dsy** sporocyte with 2 spindles. (E) Dyad of secondary sporocytes with micronuclei or lost chromosomes in the cytoplasm. (F) A polyad of microspores resulting from the desynaptic meiosis. (G) An *aph*1/*aph*1 sporocyte undergoing cytokinesis has multiple phragmoplasts. The primary phragmoplast is positioned between two daughter nuclei shown in (H), but an additional phragmoplast (arrow) is only associated with one nucleus. (I) Telophase *aph*1 cell with wavy phragmoplast. (J &K) Two focal planes of a *aph*1/*aph*1 dyad with multiple and undulating phragmoplasts. (L) Product of the *aph*1 division contains 4 nuclei and several cells without nuclei, or cytoplasts. (× 700)

pairing failure during prophase I (Fig. 2). Instead of the normal 10 bivalents or homologous chromosome pairs (Fig. 2A), *dsy*/dsy** microsporocytes have a variable number of univalents present at late prophase I (Figs. 2B, 2C). Absence of pairing leads to multiple spindles at MI (Fig. 2D). Chromosome segregation is abnormal and the products of division I usually contain several condensed chromosomes or micronuclei distributed throughout the cytoplasm (Fig. 2E). Although the second division does occur, its nature is influenced by the disruptions caused in meiosis I. The most frequent end-products of meiosis are polyads of microspores (Fig. 2F) with multiple nuclei, micronuclei, and lost chromosomes. The phenotype of this mutation is similar to other desynaptic mutations, especially *as* (Table 1), but complementation tests indicate that it is not allelic to *as*, *dsy*1, *dsy*2, or *dy* (CJ Staiger, RK Dawe, and WZ Cande, unpublished).

The second new meiotic mutant, *abnormal phragmoplast* (*aph*1), has a very exciting phenotype for workers interested in the mechanisms of plant cytokinesis (Fig. 2). During wild type microsporogenesis, the phragmoplast forms between two daughter nuclei as a proliferation of the spindle midzone which subsequently expands centrifugally until reaching the side walls. In *aph*1/*aph*1 plants cytokinesis is abnormal in 58% of all microsporocytes. Several types of disruption are apparent in these cells. Often extra phragmoplasts form at random locations in the cytoplasm demonstrating independence from nuclear position (Figs. 2G, 2H). In other cells, a single phragmoplast undergoes an uncharacteristically wavy propagation before reaching the cortex (Fig. 2I). A few cells have both phenotypes; extra phragmoplasts and distorted propagation (Figs. 2J, 2K). Successful cell plate formation by these extra phragmoplasts often results in cytoplast (enucleate cell) formation (Fig. 2L). Surprisingly, even in the absence of a nucleus cytoplasts can still form the cortical microtubule arrays typical of microspores (data not shown). Although the molecular nature of this mutation is unknown it provides at least two new insights about microtubule distribution in plant cells: 1) phragmoplast formation and distribution can be uncoupled from nuclear position and 2) formation of an interphase microtubule array does not require the presence of a nucleus.

4. Prospects and summary

Now that we have identified mutations that have interesting and stage-specific disruptions to cytoskeletal organization, serious thought must be given to pursuing these at a molecular level. One alternative to the random hunt through transposon mutagenized material described above is to directly tag the gene using the existing reference allele as a marker. This strategy has several advantages. First it allows one to pursue genes that have been deemed of considerable interest by cytological and cytoskeletal criteria. Second, it requires fewer generations to identify new transposon induced mutations. The methodology would involve crossing pollen from a line carrying an active transposable element (preferably located in a marker gene on the same chromosome arm as the gene of interest) onto ears of homozygous male sterile plants. Progeny from many such crosses are then screened for the male sterile phenotype. Any male sterile plant will be due to failure of a transposon induced event to complement the reference allele, or pollen contamination from heterozygous sibs carrying the reference allele. The latter events can be eliminated by using recessive markers closely linked to the reference allele. Based on the cytoskeletal disruptions reported here, the readily identifiable gross phenotype of the mutant, and proximity to the site of an active transposon insertion (*Ac* at *Pvv*), *ms*17 is a particularly good choice for a directed tagging approach. Future effort will be devoted to this and to the molecular characterization of those mutants isolated from *Mutator* families.

Acknowledgements

This work was supported by a USDA predoctoral fellowship (CJS) and a USDA grant (WZC). We wish to thank RK Dawe and L Dolan for critically reading the manuscript.

References

1. Albertsen MC and Phillips RL (1981) Developmental cytology of 13 genetic male sterile loci in maize. Can J Genet Cytol 23: 195–208.
2. Baker BS, Carpenter ATC, Esposito MS, Esposito RE and Sandler L (1976) The genetic control of meiosis. Ann Rev Genet 10: 53–134.
3. Baker RL and Morgan DT (1969) Control of pairing in maize and meiotic interchromosomal effects of deficiencies in chromosome 1. Genet 61: 91–106.
4. Beadle GW (1932) A gene in *Zea mays* for failure of cytokinesis during meiosis. Cytologia 3: 142–155.
5. Beadle GW (1932) Genes in maize for pollen sterility. Genet 17: 413–431.
6. Beadle GW (1933) Further studies of *Asynaptic* maize. Cytologia 4: 269–287.
7. Beadle GW (1937) Chromosome aberration and gene mutation in *sticky* chromosome plants of *Zea mays*. Cytologia Fujii Jubilee Vol, 43–56.
8. Beadle GW and McClintock B (1928) A genic disturbance of meiosis in *Zea mays*. Science 68: 433.
9. Cabral F (1989) Genetic approaches to spindle structure and function. In: Mitosis: Molecules and Mechanisms. Hyams JS, Brinkley BR (eds) pp. 273–302. San Diego, Academic Press, Inc.
10. Carlson WR (1988) The cytogenetics of corn. In: Corn and Corn Improvement. Agronomy Series. No. 18. Sprague GF, Dudley JW (eds) pp. 259–343. American Society of Agronomy, Madison, WI.
11. Carniel K (1961) Beitrage zur entwicklungsgeschichte des sporogenen gewebes der Gramineen und Cyperaceen. I. *Zea mays*. Osterreiche Bot Zeit 108: 228–237.
12. Chang MT and Neuffer MG (1989) Maize microsporogenesis. Genome 32: 232–244.
13. Clark FJ (1940) Cytogenetic studies of divergent meiotic spindle formation in *Zea mays*. Amer J Bot 27: 547–559.
14. Coe E, Neuffer G, Hoisington D and Chao S (1991) Gene list and working maps. Maize News Lett. 65: 129–164.
15. Emerson RA (1932) A recessive zygotic lethal resulting in 2:1 ratios for normal vs. male-sterile and colored vs. colorless pericarp in F2 of certain maize hybrids. Science 75: 566–567.
16. England DJ and Neuffer MG (1987) Chromosome 8 linkage studies. Maize News Letter 61: 51.
17. Golubovskaya IN (1979) Genetic control of meiosis. Int Rev Cytol 58: 247–290.
18. Golubovskaya IN (1987) Mapping of two mei-genes of maize with the help of B-A translocations. Genetika 23: 473–480.
19. Golubovskaya IN (1989) Meiosis in maize: *mei* genes and conception of genetic control of meiosis. Adv Genet 26: 149–192.
20. Golubovskaya IN and Distanova EE (1986) Locating the mei gene *ms43* in maize with B-A translocations. Genetika 22: 1173–1180.
21. Golubovskaya IN, Grebennikova ZK and Mashnenkov AS (1990) A case of simulataneous mutation of the *mei* gene arresting meiosis at prophase I and of the ms gene in corn treated with n-nitroso-n-methylurea. Genetika 26: 1249–1257
22. Golubovskaya IN, Khristolyubova NB, Urbakh VG and Safonova VT (1984) Double meiotic mutants of maize and the problem of genetic control of meiosis. Doklady Akad Nauk USSR 274: 423–427.
23. Golubovskaya IN and Mashnenkov AS (1975) Genetic control of meiosis. I. Meiotic mutation

in corn (*Zea mays* L.) *afd*, causing the elimination of the first meiotic division. Genetika 11: 810–816.

24. Golubovskaya IN and Mashnenkov AS (1976) Genetic control of meiosis communication II. A desynaptic mutant in maize induced by N-nitroso-n-methylurea. Genetika 12: 123–128.

25. Golubovskaya IN and Mashnenkov AS (1977) Multiple disturbance of meiosis in corn caused by a single recessive mutation *pam*A-A344. Genetika 13: 1278–1287.

26. Golubovskaya IN, Safonova VT and Khristolyubova NB (1980) Sequence of incorporation of meiotic genes of corn during meiosis. Doklady Akad Nauk USSR 250: 458–460.

27. Golubovskaya IN and Sitnikova DV (1980) Three meiotic mutations disturbing chromosome segregation at the first meiotic division in corn. Genetika 16: 656–666.

28. Golubovskaya IN and Urbakh VG (1981) Study of allelism of meiotic mutations with phenotypically similar disturbances of meiosis. Genetika 17: 1975–1982.

29. Greyson RI, Walden DB and Cheng PC (1980) LM, TEM, SEM observations of anther development in the genic male sterile (*ms9*) mutant of corn *Zea mays*. Can J Genet Cytol 22: 153–166.

30. Hake S, Vollbrecht E and Freeling M (1989) Cloning *Knotted*, the dominant morphological mutant in maize using *Ds2* as a transposon tag. EMBO J 8: 15–22.

31. Kaul MLH (1988) Male Sterility in Higher Plants. Wien, Springer-Verlag, 1005 pp.

32. Kuwada Y (1911). Meiosis in the pollen mother cells of *Zea mays* L. Bot Mag Tokyo 25: 163–181.

33. Lloyd CW (1987) The plant cytoskeleton: The impact of fluorescence microscopy. Ann Rev Plant Physiol 38: 119–139.

34. Mashnenkov AS and Golubovskaya IN (1980) Meiotic corn mutations induced by nitrosoalkylureas. Genetika 16: 1021-1026.

35. McCarty DR, Carson CB, Lazar M and Simonds SC (1989) Transposable element-induced mutations of the *viviparous-1* gene in maize. Dev Genet 10: 473–481.

36. McIntosh JR and Koonce MP (1989). Mitosis. Science 246: 622–628.

37. Miller OL (1963) Cytological studies in *Asynaptic* maize. Genet 48: 1445–1466.

38. Moens PB (1987) Introduction to meiosis. In: Meiosis. Moens PB (ed) pp. 1–17. Orlando, FL, Academic Press, Inc.

39. Nelson OE and Clary GB (1952) Genic control of semi-sterility in maize. An inbred with pollen semi-sterility and ovule semi-sterility caused by different genes. J Hered 43: 205–210.

40. Palmer RG (1971) Cytological studies of *Ameiotic* and normal maize with reference to premeiotic pairing. Chromosoma 35: 233–246.

41. Pan YB and Peterson PA (1989) Tagging of a maize gene involved in kernel development by an activated *Uq* transposable element. Mol Gen Genet 219: 324–327.

42. Reeves RG (1928) Partition wall formation in the pollen mother cells of *Zea mays*. Amer J Bot 15: 114–124.

43. Rhoades MM (1950) Meiosis in maize. J Hered 41: 58–67.

44. Rhoades MM and Dempsey E (1966) Induction of chromosome doubling at meiosis by the elongate gene in maize. Genet 54: 505–522.

45. Rhoades MM and Dempsey E (1985) Structural heterogeneity of chromosome 10 in races of maize and teosinte. In: Plant Genetics. Freeling M (ed) pp. 1–18. New York, Alan R Liss, Inc.

46. Schmidt RJ, Burr FA and Burr B (1987).Transposon tagging and molecular analysis of the maize regulatory locus *opaque-2*. Science 238: 960–963.

47. Schwartz D (1958) A new temperature-sensitive allele at the *sticky* locus in maize. J Hered 49: 149–152.

48. Shamina NV, Golubovskaya IN and Gruzdev AD (1981) Spindle disturbances in some meiotic mutants of maize. Tsitol 23: 275–282.

49. Shamina NV and Gruzdev AD (1987) A study of the birefringence of cell division spindle in the meiotic mutants of *Zea mays*. Tsitol 29: 104–108.

50. Staiger CJ and Schliwa M (1987) Actin localization and function in higher plant cells. Protoplasma 141: 1–12.

51. Staiger CJ and Cande WZ (1990) Microtubule distribution in *dv*, a maize meiotic mutant defective in the prophase to metaphase transition. Dev Biol 138: 231–242.

52. Staiger CJ and Cande WZ (1991) Microfilament distribution in maize meiotic mutants correlates with microtubule organization. The Plant Cell 3: 637–644.

53. Staiger CJ (1992) Indirect Immunofluorescence: The Cytoskeleton. In: The Maize Handbook. Freeling M, Walbot V (eds) New York, Springer-Verlag (in press).

54. Staiger CJ and Cande WZ (1992) *Ameiotic*, a gene that controls meiotic chromosome and cytoskeletal behaviour in maize. Dev Biol 154: 226–230.

55. West DP and Albertsen MC (1985) Three new male-sterile genes. Maize News Letter 59: 87.

13. Cell cycle regulation in cultured cells

DAVID E. HANKE

Abstract

Although plant cell cultures offer unique advantages as an experimental system for cell cycle research, would-be users should be aware of the problems of culture-induced variation, traceable to a feed-forward cycle involving competitive selection and somatic cell cytogenetics. The most useful contribution of cell cultures has been to provide material with a high degree of synchrony for cell cycle research. Investigations of the control of the mitotic cycle by plant growth substances are bogged down in the slough of signal transduction, currently an area of ignorance. There is a pressing need for more evidence of causal connection between an early biochemical event induced by the signal and signal-induced progress through the cell cycle.

There are several reasons why plant cell cultures appear to be specially convenient for the investigation of the cell cycle. First, there's no shortage of the cycles – these cells can sustain an apparently endless succession of rounds of cell division. Second, teeming heterotrophically in a chemically defined medium their environment can be tightly controlled. As a result many environmentally induced artifacts can be eliminated. Working with other types of plant cells the events following wounding, light signals and water stress for example (to all of which plants are exquisitely sensitive) may be confused with components of the cell cycle.

However we now know that these attractions are severely offset by a devastating drawback. It is this. Although you can keep their environment constant, you can't keep such complicated systems as cells constant. The problem is that the fourth dimension of time enables them to evolve, and two features ensure that they evolve rapidly. First, the cell is only required to carry out light household duties and no longer is a lethal failure if it fails to express every gene correctly, the cultured cell nucleus is free to scramble much of the genetic information, generating variants. Second, the separation of daughter cells in a suspension culture makes them into separate individuals for selection to act upon. Variants that require less raw materials, or that sequester resources away from their competitors, or that poison other cells, will replace other variants.

The pace of this evolution is accelerated by a feed-forward influence of the second feature on the first feature. Selection favours variants that complete the cell cycle in a shorter time, compressing S-phase to such an extent that some

J.C. Ormrod and D. Francis (eds.), Molecular and Cell Biology of the Plant Cell Cycle, 173–178.
© 1993 *Kluwer Academic Publishers. Printed in the Netherlands.*

late-replicating telomeric and pericentromeric heterochromatin is caught out by the early termination of S-phase and fails to be duplicated. The still unreplicated regions link sister chromatids at the following mitosis in the form of anaphase bridges which are snapped, generating busted ends that anneal around in the daughter nuclei. In the consequent welter of DNA repair, mismatch base pairing generates real mutations which mitotic crossing over subsequently exposes as homozygotes. The end result of all this is translocations, inversions, deletions and duplications, point mutations, positional effects and shifting patterns of methylation, all of which contribute to the generation of cell-to-cell variation, both genetic and epigenetic [14]. The nucleus of a cultured plant cell is, then, a genetic mill and this property is the root cause of such well known phenomena as somaclonal variation [6], and the rapid disappearance with subculturing of secondary product synthesis and regenerability. Its existence limits the usefulness of plant cell cultures since the characteristics of an individual line do not remain the same for long, and divided into separate lines their characteristics will progressively diverge. Making a virtue of necessity, we have sought to exploit it by using the somatic cell genetic mill as a correlation smasher [9].

Nevertheless, work using plant cell cultures has made recent contributions to our understanding of the control of the cell cycle. The great advantage is that they are especially useful for providing plant cells cycling in synchrony. This is because a suspension culture is the only experimental system for which added chemicals reach all cells simultaneously. Cultures treated with specific inhibitors of DNA synthesis (aphidicolin [15], hydroxyurea [18]) or mitosis (colchicine [8]) will begin to traverse the cell cycle in step after the inhibitor is simultaneously washed out of all cells. Not surprisingly cell cultures vary in the extent to which they can be synchronized, and tobacco, cultivar Bright Yellow 2, (BY-2) cells seem to give the best results, an observed mitotic index of 60 to 70% ([15], see T Nagata and Y Takahashi, this volume). Cell culture material at different cell cycle stages is used to make cDNA libraries which can be analysed by differential screen to isolate stage-specific mRNAs, e.g. that from the gene *cyc* 07, an S-phase-specific sequence from *Catharanthus* [12]. The short time scale of these experiments makes it less likely that culture-induced variation will be a problem. In future the use of subtractive hybridisation [19] should enable us to enrich cDNA libraries in differentially expressed mRNAs, increasing the haul and, therefore, the chances of isolating rarer regulatory as opposed to regulated genes.

Cultured cells are also playing a part in the elucidation of the cell cycle phase-specific cyclins and the versatile p34 protein kinase regulating the transitions between phases. A carrot A-type cyclin was under-expressed in rapidly dividing, auxin-treated cultures relative to cultures from which auxin had been withdrawn and embryogenesis was underway [10]. Both the size and relative abundance of *cdc*2-related transcripts shifted following auxin treatment of alfalfa cultures ([11], see D Dudits *et al.*, this volume).

These examples point up the characteristic feature of cell cycling in cultured cells – its dependence on hormonal signals. All cell cultures need auxin; some also require cytokinin. The chain of cause and effect linking these easily manipulable chemical signals through to progress around the mitotic cycle has been a tempting model system for cell cycle control for years and remarkably little progress has been made. We seem to be stuck at signal transduction, the black hole that blights, blocks and bars the way for so many aspects of plant development.

Because with few exceptions cultured cells need auxin for any and all movement through the cell cycle (auxin-deprived cells do not arrest at specific stages [7]) we have largely confined our attention to investigating cytokinin-alleviable blocks in the cycle. In almost all cases these are not stable. For tobacco, sycamore, carrot and other cell lines the cultures gain cytokinin-independence spontaneously after only a short time in culture. Measurements of endogenous levels of cytokinin indicate that there is no increase in the internal concentration [16], suggesting that the cytokinin-alleviable block was an insertion in the mitotic cycle and that cytokinin-independence is the default mode, achieved as the result of the loss of this insertion. The block would seem to be a luxury function whose expression is quickly lost in the competition between variants.

Soybean callus isolated from the cotyledons is pretty well unique because the requirement for cytokinin is stable. In thirty years of working with the original isolate, we have only once had a line which achieved cytokinin-independence, and no increase in cytokinin content could be detected. Other parts of the soybean seedling (and the vascular tissue of the cotyledon) yield callus that is not dependent on cytokinin, indicating that in this species dependence is a feature of the differentiated state and confirming that cytokinin-alleviable blocks are not fundamental features of the plant mitotic cycle.

We surmise that the stability of the cytokinin-dependence of cotyledonary callus is because there are a number of cytokinin-alleviable blocks; each at different stages of the cell cycle. In the random 'mill' of the cultured cell nucleus, the likelihood that one of these may cease to be expressed is appreciable, but the chances that all of them will be suppressed are slender. There is evidence for multiple blocks in this tissue from a study of the fluctuation in mitotic index on supplying cytokinin. Cell division recommenced in a series of pulses, each involving about one quarter of the cells [21]. Recently, we were able to confirm this from the results of an immunohistochemical study of the pattern of resumption of S-phase.

Because of the unique stability of the response we chose cotyledonary callus of soybean for an investigation of the transduction process, the chain of cause and effect that links the cytokinin signal to the induction of cell division. Because a number of calluses require inositol to sustain cell division [1], and in one case to respond fully to cytokinin [2], we began by looking for cytokinin-induced alterations in phosphatidylinositol turnover and found persistent and clear effects of cytokinin treatment [4]. Five years later, phosphatidylinositol was still turning over at the same rate but was no longer affected by cytokinin-treatment [9]. The genetic mill of the cultured cell nucleus had apparently broken the correlation. It seems that the cell culture from cotyledonary soybean callus can lose the link between cytokinin and phosphatidylinositol turnover while retaining the link between cytokinin and the cell division cycle.

There are at least two other early, biochemical effects of cytokinin in these same soybean cells that are apparently not related to its effect on the mitotic cycle, even though they conform to "classic" signal transduction phenomena. First, cytokinin-induced recruitment of ribosomal subunits into polysomes was only found when cytokinin was added at the same time as fresh growth medium. The effect of cytokinin alone was to induce division without the increase in polysomes [3]. Second, although a 15 mV hyperpolarisation of the plasma membrane due to stimulation of the H^+- pump took place over the first 2 min after supplying 2 μM cytokinin, 2 μM adenine (inactive for cell division) elicited the same effect, and two active but N-9 substituted cytokinins did not [17].

Two alternative explanations can be suggested.

The first is that many of the early events triggered by signals are not connected to the subsequent developmental response in question. It is clear that cytokinins can cause a wide range of different, early, biochemical affects: on phosphatidylinositol turnover, on H^+- pumping and on protein synthesis, but each of these can be shown to be unnecessary for relieving the cell cycle block. The possibility of gratuitous association between the early biochemical events triggered by growth substances and the downstream physiological and developmental consequences of growth substance treatment should be more widely acknowledged. A correlation is not a causal connection. Work with G1-arrested cultured cells of *Catharanthus* that links auxin stimulation of putative inositol-1,4,5-trisphosphate release from phosphatidylinositol-4,5-bis-phosphate at 1 min with auxin-induced cell division by 24 h [5] is still at the correlation stage.

The second explanation is that there is a redundancy of signal transduction mechanisms. The same signal may be causally connected to the same response via several different and separate mechanisms, providing alternative modes of transduction. Examined individually all of the mechanisms will appear to be unnecessary for the final response. There is some support for this idea from work on the role of Ca^{2+} in the signal transduction mechanism mediating abscisic acid-induced stomatal closure. Artificially increasing the level of cytosolic Ca^{2+} consistently leads to closure. Treatment with abscisic acid only increased the concentration of Ca^{2+} in the cytoplasm in a minority of cells but induced closure in all of them [20]. Certainly, a redundancy of signals is widely observed in the control of individual responses by plants. If redundancy of transduction mechanisms turns out to be a widespread feature, establishing causality will require a fundamental change in approach. Examined individually all of the mechanisms will appear to be unnecessary for the final response. Only if every single avenue is blocked can function be established, a tall order.

As yet we are not even in a position to decide which of the two alternative explanations is most likely. For the future, only genetic strategies, such as a positive screen for mutants defective in a signal transduction pathway [13], offer any glimmer of hope. To understand signal-dependent progress through the cell cycle in cultured cells is a tough challenge, with little prospect of early success.

Acknowledgements

Our work on the control of the cell division cycle in soybean cell cultures by cytokinin was supported by a research grant from AFRC.

References

1. Anderson L and Wolter KE (1966) Cyclitols in plants: biochemistry and physiology. Annu Rev Plant Physiol 17: 209–222.
2. Bender L and Neumann K-H (1978) Investigations on the influence of pre-culture in IAA- and kinetin-containing culture media on subsequent growth of cultured carrot explants. Zeits Pflanzenphysiol 88: 201–208.
3. Bevan M and Northcote DH (1981) Subculture-induced protein synthesis in tissue cultures of *Glycine max* and *Phaseolus vulgaris*. Planta 152: 24–31.
4. Connett RJA and Hanke DE (1987) Changes in the pattern of phospholipid synthesis during the induction by cytokinin of cell division in soybean suspension cultures. Planta 170: 161–167.
5. Ettlinger C and Lehle L (1988) Auxin induces rapid changes in phosphatidylinositol metabolites. Nature 331: 176–178.
6. Evans DA (1989) Somaclonal Variation – genetic basis and breeding applications. Trends Genet 5: 46–50.
7. Everett NP, Wang TL, Gould AR and Street HE (1981) Control of the cell cycle in cultured plant cells II. Effects of 2,4-D. Protoplasma 106: 15–22.
8. Guri, A, Zelcer A and Izhar S (1984) Induction of high mitotic index in *Petunia* suspension cultures by sequential treatment with aphidicolin and colchicine. Plant Cell Reports 3: 219–221.
9. Hanke DE, Davies, H, Biffen M, Connett RJA and Freathy TC (1990) Cytokinin mode of action-problems and perspectives. In: Plant Growth Substances (1988) RP Pharis and SB Rood (eds) pp. 161–172. Berlin: Springer-Verlag.
10. Hata S, Kouchi H, Suzuka I and Ishii T (1991) Isolation and characterization of cDNA clones for plant cyclins. EMBO J 10: 2681–2688.
11. Hirt H, Pay A, Gyorgyey J, Bako L, Nemeth K, Bogre L, Schweyen RJ, Heberle-Bors E and Dudits D (1991) Complementation of a yeast cell cycle mutant by an alfalfa cDNA encoding a protein kinase homologous to p34^{cdc2}. Proc Natl Acad Sci USA 88: 1636–1640.
12. Ito M, Kodama H and Komamine A (1991) Identification of a novel S-phase-specific gene during the cell cycle in synchronous cultures of *Catharanthus roseus* cells. Plant J 1: 141–148.
13. Karlin-Neumann GA, Brusslan JA and Tobin EM (1991) Phytochrome control of the *tms*2 gene in transgenic *Arabidopsis*: A strategy for selecting mutants in the signal transduction pathway. Plant Cell 3: 573–582.
14. Larkin PJ, Banks PM, Bhati R, Brettell RIS, Davies PA, Ryan SA, Scowcroft WR, Spindler LH and Tanner GJ (1989) From somatic variation to variant plants: mechanisms and applications. Genome 31: 705–711.
15. Nagata T, Okada K and Takebe I (1982) Mitotic protoplasts and their infection with tobacco

mosaic virus RNA encapsulated in liposomes. Plant Cell Reports 1: 250–252.

16. Nakajima H, Yokota T, Matsumoto T, Noguchi M and Takahashi N (1979) Relationship between hormone content and autonomy in various autonomous tobacco cells cultured in suspension. Plant Cell Physiol 29: 1489–1499.

17. Parsons A, Blackford S and Sanders D (1989) Kinetin-induced stimulation of electrogenic pumping in soybean suspension cultures is unrelated to signal transduction. Planta 178: 215–222.

18. Pfosser M (1989) Improved method for critical comparison of cell cycle data of asynchronously dividing and synchronized cell cultures of *Nicotiana tabacum*. J Plant Physiol 134: 741–745.

19. Timblin C, Battey J and Kuehl WM (1990) Application of PCR technology to subtractive cDNA cloning. Nucleic Acids Res 6: 1587–1593.

20. Trewavas A and Gilroy S (1991) Signal transduction in plant cells. Trends Genet 7: 356–361.

21. Wang TL, Everett NP, Gould AR and Street HE (1981) Control of the cell cycle in cultured plant cells III. Effects of cytokinin. Protoplasma 106: 23–26.

14. The cell division cycle in relation to root organogenesis

PETER W. BARLOW

Abstract

Cell division in plants is one component of the process of organogenesis. In the case of roots, division can be viewed from two perspectives, one relating to its structural role in blocking out the cellularized pattern of the organ, the other emphasising its functional importance in supplying cells for growth. In neither case is division directly relevant to tissue differentiation since this probably results from positional cues superimposed on the cellularized whole; new cells created by division are, however, the units in which differentiation is accomplished. The structural aspect of division in relation to organogenesis emphasises the orientation of the new cell walls in various regions of the meristem. It also recognises two basic classes of division, the formative and the proliferative, properties of which are illustrated with examples from tomato and maize roots, respectively. Formative divisions occur in a programmed sequence which has been worked out for the cortex and the cap/dermatogen cell complexes. Programmes also govern the proliferative divisions and details are given of two of these for stele and cortex. Since the division sequences are recursive, they are amenable to analysis by means of L-systems. These afford an opportunity to formalize the portion of the epigenetic code that applies to cell patterning. At the deeper, cytological level, the code may resolve into recursive patterns of microtubule behaviour.

1. Introduction

The root apical meristem is excellent material for studying cell division in plants. Not only is it a rich source of dividing cells which are indispensible for exploring the molecular controls of the mitotic cycle, but it is also a site of organogenesis. In this latter respect, it lends itself to investigations of how the mitotic cycle is organized, both spatially and temporally, and how it is associated with specific patterns of cellular differentiation.

Molecular activities that provide the 'motive force' for the mitotic cycle do not operate autonomously within the root apex. They are permitted to function only by virtue of a supportive physiological milieu. The ultimate regulator of mitotic activity in multicellular plants is, however, the external environment. Energy from the environment is transduced into a multiplicity of informational

179

J.C. Ormrod and D. Francis (eds.), Molecular and Cell Biology of the Plant Cell Cycle, 179–199.
© 1993 *Kluwer Academic Publishers. Printed in the Netherlands.*

signals which filter through many levels of cellular organization before being received by potentially meristematic cells. For example, radiation (photons) and chemical energy are converted into long-range correlative information in the form of hormonal and nutritional signals, while additional and more locally acting sources of mitotic-regulating information are received from neighbouring cells in the form of molecules which cross plasmodesmata and biophysical signals such as stress patterns and electrical currents. This holistic scheme of cell cycle regulation is summarised in Fig. 1.

Other articles in this volume deal with molecular aspects of mitotic regulation in plants at the level of the cell. There, the problem is the means by which one cell begets another. The present article is concerned with the cell cycle in relation to organogenesis and morphogenesis. Here, the focus is on how patterns of cell division are propagated through many cycles, thus creating, as well as perpetuating, the form of an organ – in this case the root. It is now axiomatic that molecular activities are determined by a self-reproducing genetic code based on DNA. This is why mutational analysis of cycle regulation, at the level of biochemical pathways, has been so successful in fission yeast, for

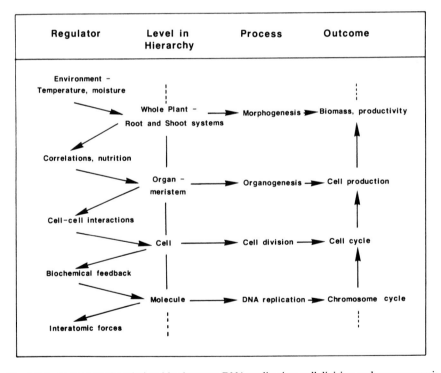

Fig. 1. Scheme to show the relationships between DNA replication, cell division and organogenesis in a plant, and also the factors that regulate these processes. The whole is an autopoietic system [35] where the sequential outcome of processes (horizontal arrows) engender, by their addition to the complexity of the system (upward-directed arrows), a cascade of regulatory factors (downward-directed arrows).

example [10, 28], and why these simple organisms serve as models for cycle regulation in eukaryotes (see J Hayles and P Nurse, this volume). However, the perpetuation of division patterns (or pathways) and the development of organic form also require a self-reproducing source of information which supplements that contained in the genome. This is referred to here as epigenetic information and is assumed to be based on a code located somewhere in the structure of the cell itself [5] or in the cell's interactions with its neighbours [24]. The nature of this information is much less well understood than that of the DNA code, but root meristems are one useful place to search for clues concerning its properties.

2. Cell division in the root apex

Root apices of higher plants are simple organogenetic systems which grow axially. Unlike shoot apices, there are no cyclical re-arrangements of division patterns such as are associated with the inception of leaves (see D Francis and RJ Herbert, this volume). Although roots do form lateral organs, these are not usually initiated within the apex itself, but much further back. (Some ferns are exceptional; their lateral root primordia can be formed as part of the normal sequence of divisions within the apical meristem.) Divisions in the apex can be classified according to either their structural or their functional attributes.

2.1. Structural and functional classifications

Two types of division, formative and proliferative [16], are recognizable on account of the structured pattern of cells they create. Formative divisions are those which generate new files of cells and are usually located at the tip of the root. At cytokinesis, cells divide with orientations that are either radial-longitudinal or circumferential (periclinal) with respect to the root and/or cell-file axis. The sites of such divisions can be deduced in longitudinal sections from the presence of T or ⊥ wall junctions (cf. Fig. 5) where one longitudinal file of cells has branched into two files. Proliferative divisions are transverse to the file axis and have the effect of increasing the number of cells in the files created by formative divisions. They may be interspersed with formative divisions within the apex, but mostly they occur proximal to this zone and thus comprise the majority of divisions within the meristem.

Most of the cell-files of roots with closed meristems can be traced to a few cells at the tip. These cells can be termed 'structural initials' to emphasise their structural role in organogenesis; they comprise the 'minimal constructional centre' in Clowes' terminology [11]. However, it should be recalled that such initials are not permanent cells, nor do they necessarily have a permanent location since new initials are continually being created by division. The zone housing the structural initials is known as the quiescent centre (QC); many formative divisions occur within the QC.

Divisions in the root apex can also be classified according to their functional properties. Considering the root proper (i.e., ignoring the root cap), a small group of cells at its extreme tip shows low rates of division and possesses many properties similar to 'stem-cell' compartments in the proliferative systems of animals [30]. This is the QC referred to above. Proximal to the QC is the meristem of the root proper which consists of divisions that amplify the number of cells derived from the QC. The rapidly cycling cells at the boundary between the stem-cell/QC compartment and the amplification compartment may be considered as 'functional initial' cells.

2.2. Cell kinetics and cell differentiation in relation to cell division

Cells divide a variable number of times depending on their location within the meristem. Coincident with the cessation of division, the cells enter a compartment of accelerating elongation. When this process is completed, the cells mature. There are indications that the rate of flow of cells into and out of each compartment is regulated homeostatically, which means that the rate of cell production by the meristem is subject to an overall control determined by the state of the root as a whole (cf. Fig. 1). Probably, the factors governing the dynamics of cells in the apex are no less complex than those of self-maintaining animal systems such as blood tissue [36].

Running in parallel with meristematic activity is a process of differentiation whereby cells in different files develop as epidermis, cortex, etc. Differentiation has some interaction with division since the various types of cell are associated with the different numbers of divisions in the amplification compartment: cells of the xylem and phloem lineages undergo the fewest divisions, pericycle the most. Cells at the extreme tip of the root differentiate as cap. Here, the meristem is short (as is the cap itself owing to the constant loss of cells from its surface), and most of the cells are displaced forwards towards the tip rather than backwards away from the tip as in the meristem of the root proper. Those that are displaced basipetally and adhere to the epidermis form a dermatogen.

3. Experimental systems

The relationship between the mitotic cycle and organogenesis of roots has been studied in two species. Isolated roots of tomato (*Lycopersicon esculentum* cv. Ailsa Craig and cv. Moneymaker) grown *in vitro* have a regular pattern of formative divisions which can be manipulated by both genetic and nutritional factors [6, 7]. The periclinal class of divisions can be studied most easily in median longitudinal sections (LS); transverse sections (TS) are necessary to record the sequence of longitudinal-radial divisions. Proliferative divisions have been investigated in seedling roots of maize (*Zea mays* cv. LG11) grown in vermiculite. These divisions also show distinctive and reproducible patterns which can be investigated experimentally [3, 4]. In both systems, it seems that the division patterns are self-perpetuating and hence are determinate. The duration of the cell cycle and its phases are also known in these roots [8], but these parameters are relevant only in so far as they relate to the rate at which the organogenetic pattern unfolds since, as far as is known, no aspect of the cell cycle (except for cytokinesis) determines the pattern itself.

Patterns of division are difficult to describe without some suitable notation and, in this respect, the application of L-systems is useful [5]. At present, and as will be described below, L-systems have been applied only to the proliferative divisions of maize [26], but progress is being made (e.g. 27) to utilize them for the more complex formative divisions of tomato.

4. Formative divisions of tomato roots

4.1. The cortex

When seen in LS, each of the cell files that run the length of the cortex traces to a file mother cell located in the innermost row of the cortex (Fig. 2). The proximal transverse wall of this cell forms the base of the ⊥ junction of walls where two other cell-files meet. Ultimately, all files seen in LS reduce to one or two tiers of cells at the pole of the root (i.e., they trace to structural initials within the QC) sandwiched between the cap meristem and stele. The ⊥ junction results from a periclinal division in a file mother cell during a previous cell generation. The mother cells also divide longitudinal-radially and transversely. The probability with which each class of division occurs depends on the location of the mother cell within the inner cortical file: cells at, or close to, its summit

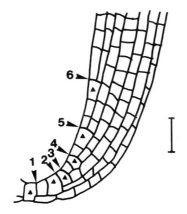

Fig. 2. Cellular patterns in the cortex of a 7-day-old tomato root grown *in vitro* (with 1.5% sucrose) seen in median longitudinal section. Arrowheads point to the ⊥ cell wall junctions which are numbered according to their position along the length of the innermost row of cortex; solid triangles indicate the mother cells of the cortical cell-files. The three mother cells at the distal end of the cell system, where the files converge, would be housed within the quiescent centre. Scale bar = 25 μm.

have equal probabilities of dividing in all three planes. Here, they are increasing (i) the number of cells within the inner ring of cortical files by radial divisions, (ii) the number of rows by periclinal divisions, and (iii) are contributing new cells along the files by transverse divisions. More proximally, where the inner ring is complete, only periclinal divisions co-exist with transverse divisions. Eventually, when all the rows of the cortex have been formed, the periclinals cease, leaving only the transverse (proliferative) divisions.

The number of formative divisions can be modulated by the sucrose content of the medium in which the roots are grown (Fig. 3) [7]. At the threshold concentration of 0.5% sucrose, roots develop a minimum number (4) of cortical rows. Increasing the concentration to 2% causes more (maximum of 8) rows. At each sucrose concentration the number of rows, as well as the sites of the file mother cells, are predictable (Fig. 3). Interestingly, as the sucrose content is raised, the size of the QC also increases. It may be because of this that the number of formative divisions, and hence the number of cortical rows, increases, the effect being due to the predominantly isodiametric growth which occurs in the QC with the attendant lack of bias in the orientation of division. The slow rate of division in the QC may also be a means of ensuring that any epigenetic programme which specifies the division pattern, and involves a

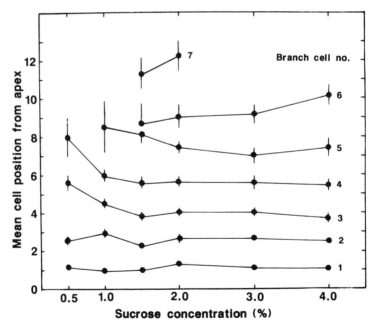

Fig. 3. Mean position of file mother cells in the innermost row of the cortex of 7-day-old tomato roots grown *in vitro* with different concentrations of sucrose. Each mother cell forms, by periclinal division, one of the rows of the cortex. These cells are associated with a ⊥ wall junction and are given a number as shown in Fig. 2 (branch 1 is the most distal, branch 7 the most proximal). Each concentration of sucrose supports a characteristic number of cortical rows with mother cells (branch positions) in predictable positions. Vertical bars = SE of mean.

response of the cell to its dimensions, is executed with maximum efficiency. An example of an inefficient system, the root apex of the *gib*-1 mutant tomato, is described later.

The accomplishment of periclinal and radial divisions in the inner cortical row is associated with a characteristic pattern of cell growth. A simple hypothesis was proposed [7] whereby transverse and longitudinal divisions are determined by the shape, or aspect ratio (A, where A = cell length/cell breadth), of a cell as it prepares for mitosis. If a critical value of A is exceeded

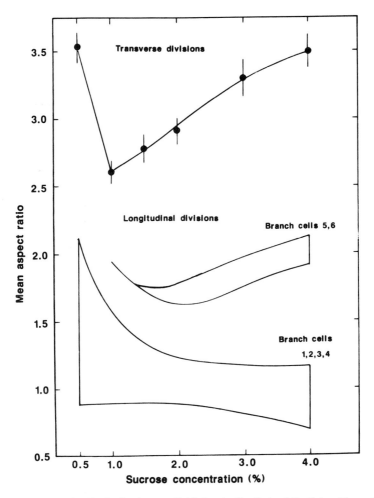

Fig. 4. Mean aspect ratio (A) of cells about to divide longitudinally (periclinally) and hence branch to give new rows from the inner cortex of tomato roots grown for 7 days *in vitro* with different concentrations of sucrose. For clarity, values of A for cells at two sets of branch position (1–4 and 5–6) are indicated as being contained within the two 'window' areas. Other cells in the inner row of the cortex, not at these branch positions, are assumed to be going to divide transversely. Vertical bars = SE of mean.

the cell divides transversely, otherwise the cell divides longitudinally. Measurements of the dimensions of large interphase cells which were obviously close to mitosis, indicated that file mother cells had values of A quite distinct from those of other cells in the same file (Fig. 4). Naturally, the shape of a cell depends on whether it has just divided or is just about to divide but, nevertheless, using the critical aspect ratio hypothesis, the relative rate of cell expansion in the longitudinal and radial planes at a given position within the file, coupled with a particular rate of division, will determine the relative frequency of longitudinal and transverse divisions at that location. A simple simulation based on these parameters confirmed this and reproduced the patterns found [7].

A more complex treatment involving a growth tensor [17] for all the cortical cells at the apex should be able to generate the complete pattern of formative and proliferative divisions in this region. In this model, the tensor is a measure of the relative rate of growth in each of the three planes. The plane of division is then governed by some simple rules associated with the dimensions of the cell (e.g., the new wall forms across a minimum area between opposite walls and transverse to the principal direction of growth – Errera's and Hofmeister's rules (see [25]). However, the tensor itself has to be determined in some way. It might, for example, partly reflect the flux of some growth regulating substance(s) (auxin, sucrose, e.g.) within the apex. This, in turn, might determine the rate of cell volume increase. Then, specific configurations of microtubules which, by governing the deposition of wall microfibrils, could dictate the major direction of cell growth. Should microtubules be subject to a programme which alters their orientation in successive cell generations (and such rotating systems of microtubules and microfibrils are known, or inferred, in various plant species [31, 32]), specific and deterministic patterns of growth and division would result. This particular aspect of cortical file formation in tomato roots is currently being investigated in conjunction with an analysis of the 3-dimensional pattern of division. Preliminary results show that microtubule orientation in file mother cells is either transverse or random (longitudinal orientations are infrequent) and that the frequency of these categories differs depending on the location of the cells. A similar programming of microtubule orientations is implicit in Lindenmayer's deterministic analysis [23] of formative divisions in the *Azolla* root.

In the apical cell of the *Azolla* root, the division plane shifts through 120° in each cycle. It might not be premature to suggest that something similar occurs in the QC zone of higher plants. This leads to the proposal of the 'tumbling helix' hypothesis for microtubules at the apex of the root. In the zone comprised of formative divisions, the orientation (pitch) of the intracellular helix of microtubules is not fixed but slowly tumbles within the cell. After each division, a new pitch to the helix is established and cell growth reorients accordingly. Cells then divide in a sequence reflecting the helical tumbling, this being manifested in the three division planes – transverse, periclinal, and radial. Away from the tip, variation in the pitch of microtubular helix becomes more restricted, that which is conducive to radial division being the first to diappear, then the periclinal, leaving finally only the orientation that permits transverse divisions.

The *gib*-1 mutation in tomato affects gibberellin biosynthesis and causes excised roots to grow more slowly than wild type *in vitro* [9]. Anatomical evidence suggests that gibberellins are required for normal orientation of cell growth in the meristem and elsewhere in the root [9]. The mutation has a profound effect on formative divisions [6]. Although *gib*-1 roots can have the same number of cortical rows as wild type, they are formed differently (Fig. 5; cf. Fig. 2). Most rows trace to an unusually large number of tiers between stele and cap meristem; only one or two rows originate in the normal way from mother cells within the inner cortical file. Exactly how the mutation brings about this altered pattern of cells is unclear. The additional tiers result from more rapid cell division in the QC (mitotic rate is, however, not affected in the proximal meristem) [9], but this does not necessarily explain why the division pattern has departed from normal. In effect, within the distal cortical cells, periclinal divisions are the rule, and radial and transverse divisions have been suppressed. Moreover, despite the breakdown of the wild type division pattern, it seems as though there must be some exchange of information within the apex that continues to maintain the usual relationship between the number of cortical rows and the sucrose concentration of the medium.

Although it was proposed above that the additional divisions in the mutant QC are periclinal with respect to the cortex/stele boundary, these divisions could also be regarded as transverse with respect to the root axis. That is, they could be interpreted as proliferative divisions belonging to an incipient cap

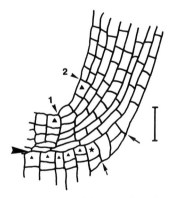

Fig. 5. Cellular patterns in the cortex and part of the cap of a 4-day-old *gib*-1 mutant tomato root grown *in vitro* (with 1.5% sucrose). Only two positions, associated with ⊥ wall junctions, where periclinal divisions increase the number of cortical rows can be seen (numbered arrowheads); other cortical rows trace to tiers of cells at the summit of the root. (Compare with the single tier at the summit of a wild-type root shown in Fig. 2.) and large solid triangles indicate initials for the cap columella. The star indicates an initial for a dermatogen cell-complex which runs up the flank of the root. Arrows point to sites of periclinal divisions that have created T wall junctions in the dermatogen. Large arrowhead at the left indicates the boundary between root and root cap. Scale bar = 25 μm.

meristem. As such, they would account for the partially 'open' appearance of the apex, even though a distinct boundary between cap and root is still present (Fig. 5). This boundary later becomes irrelevant, however, since a new cap meristem does indeed form from some of the cortical tiers within the QC of the *gib*-1 mutant root ([6], and PW Barlow, unpublished).

Gibberellins are known to stabilize microtubule arrangements in cells [28, 33]. In the *gib*-1 mutant, with its reduced gibberellin levels, the microtubules may be less stable. Consequently, they may be unable to resist rearrangement from their usual seemingly random orientation to one that is transverse with respect to the root axis and hence in conformity with, and perhaps under the influence of, cells of the cap meristem. Such an inductive influence from the cap on microtubules in cells of the QC might cause them not only to grow longitudinally but even to divide more rapidly.

4.2. Cap and dermatogen

The other location of formative divisions in the tomato root is at the junction between the meristematic initials of the central core, or columella, of the root cap and a surrounding annulus of meristematic cells (Fig. 6). Descendents of the

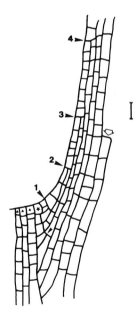

Fig. 6. Cellular patterns in the dermatogen and columella of a 4-day-old tomato root grown *in vitro* (with 1.5% sucrose). Arrowheads indicate the T wall junctions in the dermatogen (numbered according to their position with respect to the origin of the file – cf. Fig. 2), the star indicates the site of a dermatogen initial. The latter abuts columella initials (solid triangles). Sometimes a second periclinal division occurs in one of the distal descendents of a dermatogen initial cell (arrowhead within the flank of the cap). Open arrow at the right indicates a cell-file of dermatogen whose proximal portion has sloughed off from the root surface. Scale bar = 25 μm.

latter are displaced, in both a radial and a distal direction to form the flank of the cap and, in a proximal direction, the concentric layers (rows) of cells which become the dermatogen that ensheathes the surface of the root. The site of the dermatogen initials can be located by tracing to their origin the successive generations of cell complexes, each with a T wall junction at its proximal end (Fig. 6).

Only the first one or two cell-tiers of the cap columella divide (i.e., there is one tier of functional initials plus one additional tier in the amplification compartment) and do so in a predominantly transverse plane. The occasional longitudinal division admits new initials not only to the columella but also to the dermatogen. At the edge of the columella, cell growth increases in the radial direction. This abrupt switch in the orientation of growth establishes the dermatogen initials. The initials divide both periclinally and transversely to generate new sets of T wall junctions. Their positions with respect to each other lead to deductions about the rate of development of the cell complexes which they define.

The rhythm of division in the dermatogen initial is such that periclinal divisions, which form new rows, occur at nearly the same rate as the transverse

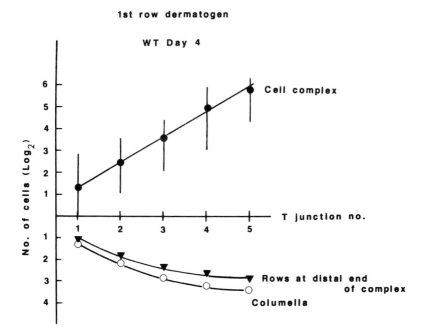

Fig. 7. Mean numbers of cells in the innermost row of dermatogen cell complexes lying between the initial cell (starred in Fig. 6) and successive T junctions (indicated on the horizontal axis; these T junctions are numbered in Fig. 6). The range of cell numbers found is indicated by horizontal bars. The line was fitted by regression. In the lower part of the graph are the number of rows at the distal end of the dermatogen complex where they abut the cap columella, and the number of contiguous columella cells. Cell numbers are given on a $\log_2$ scale to indicate cell doublings (1, 2, 4, ...).

190

divisions along each row. Thus, the number of cells intervening between initial and each successive T junction (when seen in LS) conforms approximately (but not exactly) to a doubling series (1, 2, 4, ...) (Fig. 7). The dermatogen cell-complexes, bounded at their proximal end by the capital of the T, abut the columella at their distal end. The number of these columella cells associated with each successive dermatogen cell-complex shows that meristematic columella cells divide transversely at a slightly greater rate than the dermatogen initials divide periclinally (Fig. 7). When radial-longitudinal divisions are considered, the final number of cells within the inner row of dermatogen

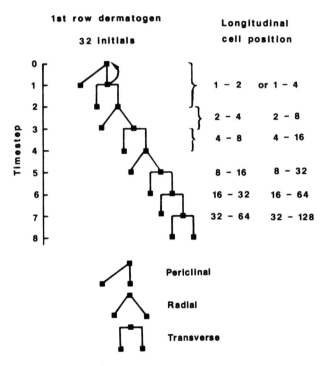

Fig. 8. Cell genealogy which describes the generation of dermatogen cell complexes and accounts for the numbers of cells found both longitudinally and radially at different distances along the length of 4-day-old, *in-vitro*-grown tomato roots (with 1.5% sucrose). The different types of division (periclinal, radial, transverse) are indicated. Only descendents of the initial cell within the inner row of dermatogen are shown. The other set of cells would also contain transverse and radial divisions, and also the occasional periclinal division (arrowed in Fig. 6). The cell numbers observed in the three dimensions of the complex are best described by the alternating activities of a pair of initial cells, consisting of a single cell and a doublet of cells. Such a scheme accounts for the variation in cell numbers found along the length of the complex (cf. Fig. 7); after the various types of division, the single initial cell would have generated the number of cells recorded in the left-hand column, the doublet of initials those in the right-hand column. In all, there are 32 initials (or 16 pairs) surrounding the cap columella. The divisions also can be matched to an actual timescale from knowledge of the doubling time of the cell population. In this case, one timestep is equivalent to about 40 h for most of the proliferative divisions [6], but may be somewhat less for the earliest formative divisions.

indicates that cells derived from each initial divide about 1.5 times in this plane. Therefore, one radial division probably follows each periclinal division (as in the inner cortical file), and a second radial division may occur after an intervening transverse division (Fig. 8). The radial divisions, unseen in LS, account for the slower increase in the number of rows of dermatogen compared to the increase in the number of cells in the associated columella, even though their interdivisional periods are similar [6] (Fig. 7).

4.3. Sequential activity of initials

In a 4-day-old tomato root grown in 1.5% sucrose, 32 dermatogen initials surround the cap columella, six initials provide for the cortical lineages, and a similar number may obtain for the vascular cylinder. Given the special significance of these initial cells, both individually for the generation of tissues and collectively for root organogenesis, it is important to know whether they have a specially organized and co-ordinated pattern of division or not.

It is difficult to answer this question precisely because it is impossible to observe the activities of the initial cells directly. The best that can be done is to make inferences from the pattern of T wall junctions in dermatogen and cortex after reconstruction of serial sections. In dermatogen, it is possible to identify T junctions in successive longitudinal sections and see that they apparently wind spirally up the root suggesting that a wave of periclinal division circulates around the ring of initial cells. However, it is difficult to establish this with certainty since not all sections are informative. (A circular succession of periclinal division could be represented in Fig. 8 as a staggering, in time, of the origins of the genealogical tree for each successive initial.) Radial divisions may also be organised spirally, as seems to be shown by the sequence of the edges of packets boundaries in reconstructions of transverse sections. The periclinal divisions of the cortex are more difficult to analyse as they are bunched together at the apex. Radial divisions, however, occur over a longer distance, and can be analysed from serial transverse sections. Again, successive branch points within packets of cells, marking sites of former radial division, may follow a spiral pattern. Spiral division patterns have a precedent in the roots of leptosporangiate ferns (e.g., *Azolla* [16]). Here, a single apical cell divides in spiral sequence; the result is a spiral of derivatives each with its own programme of divisions.

A rotational sequence of division in the initial cell has the property of harmonising the production of new cell files with rectilinear root growth. It means that chance inequalities of cell division rate within the initial zone cannot easily cause one region of the meristem to gain precedence over another with respect to cell production and the provision of cells to the elongation zone. No specific 'organizer' is required for its operation: a central cell with a rotating plane of division may be sufficient to provide the necessary mechanism (e.g., the most distal cell of the lineage shown in Fig. 2; see also Fig. 2 in [5] and ref. [27]).

5. Proliferative divisions in the maize root

5.1. Cell packets and deterministic division sequences

That programming of transverse division occurs in the proliferative portion of the meristem might be dismissed as unlikely on the grounds that this region exists simply to amplify the cell numbers required for the maintenance of rapid root growth and the supply of specialized cells to sustain the ever-expanding shoot system. Besides, for programming of division to occur, the plant must invest in a cellular apparatus which has a level of complexity greater than that of the unstructured 'default' division system. The latter system is adequate to generate divisions, but would be unable to regulate their sequence. Observations of transverse divisions in roots [1, 3, 18] do, however, reveal unequal-sized sister cells with frequencies (cf. Fig. 12) that challenge the supposition that they reflect a certain statistical distribution of the otherwise symmetric positioning of new cell walls at cytokinesis. A detailed examination of proliferative divisions in young primary roots of *Zea mays* confirms that there are indeed organized sequences of divisions [26]. These are revealed in the regular distributional patterns of sister-cell lengths that arise at successive generations.

Patterns of proliferative divisions are recognizable because cells of common ancestry remain together within cellular 'packets'. The transverse end walls of the packets were originally present in the dormant embryonic radicle. Because they are the oldest walls, they are thicker and more densely stained than any other transverse wall. On the other hand, the most recently formed walls are the thinnest and least densely stained. Thus, cells of different generations (daughters, grand-daughters, etc.) can be identified by examining the set of walls that divide up a packet. By sampling roots of different age, it is possible to find that the most recently divided cells within the packets occupy a number of alternative sites. By estimating the frequencies of these alternative sites, various sequences, or pathways (P), of division have been established [26]. In the heart of the cortex (cells originally 300 μm from the root cap junction of the dormant radicle), the two most frequently occurring pathways are P22 and P21 (terminology from [26]), while P3 is most frequent in the adjoining region of the stele (Fig. 9a, b). P22 shows a basipetal sequence of division, whereas that of P3 is largely basifugal. These pathways operate over the first three cell generations following germination and increase the cell number in the packet from 1 to 8. The next generation has been analysed in the cortex (8 cells increasing to 16 cells) and the most frequent pathways also have basipetal division sequences. The pathways are believed to persist in the respective zones of the root whatever its age since germination.

The various pathways can be simulated by means of an L-system algorithm. (L-systems are biologically motivated devices to generate sequences of state symbols.) The algorithm determines the interdivisional period (lifespan) of each cell within the growing packets. For P22, the algorithm is notated (6,5,5,4); for

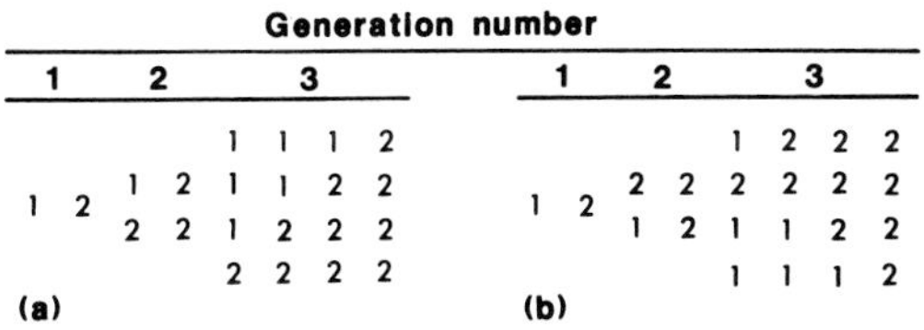
Generation number
1 2 3 1 2 3
1 1 1 2
1 2 1 1 2 2
1 2
2 2 1 2 2 2
2 2 2 2
(a)
1 2 2 2
2 2 2 2 2 2
1 2
1 2 1 1 2 2
1 1 1 2
(b)

Fig. 9. Two schemes, each giving an example of the development of cellular packets (by transverse and proliferative division), in two regions, cortex (a) and stele (b), of the primary root meristem of maize. The number of cells in the packets increases from 1 to 8 as a result of three rounds of division. In each generation, a cell is either undivided (1) or divided (2). Reading the scheme from left to right indicates the temporal evolution of the packet, and reading it from bottom to top indicates the apical-basal orientation of the packet within the root. These particular packets conform to division pathways P22 (a) and P3 (b). Other pathways have other spatial sequences of division: e.g., P21 (mentioned in text) substitutes the packet (2122) for (1222) in the penultimate position in (a).

P3, it is (4,3,4,5). The numbers in parentheses refer to the relative lifespans of cousin cells in a packet comprised of a quartet of cells. The underlined numbers indicate the lifespans of the daughters of the original mother cell. The left-to-right order of the numbers reflects the basico-apical orientation of the cells, to which these lifespans refer, within the packet with respect to the root axis. Thus, the most basal cell of a quartet of cells has the longest lifespan (6 units) in P22, whereas it is the apical cell which has the longest life span (5 units) in P3. The cells switch their lifespan at each generation according to a particular set of transition rules which can be schematised as shown in Fig. 10 using *a* and *b* to denote the apical and basal cells of a pair. The subscript numbers denote either the states of the cells or the timesteps which they are committed to traverse before dividing. The switching of the lifespans is indicated by the pair of

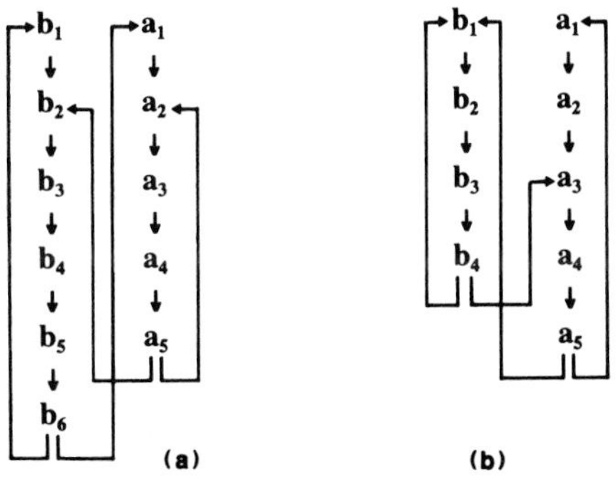
b1
b2
b3
b4
b5
b6
a1
a2
a3
a4
a5
(a)
b1
b2
b3
b4
a1
a2
a3
a4
a5
(b)

Fig. 10. Recurrence graphs of two bootstrap L-systems that describe (a) division pathway P22 in the cortex, and (b) pathway P3 found in the stele of the maize root meristem (cf. Fig. 9). The graphs relate to groups of four cousin cells with lifespans (6,5,5,4) and (4,3,4,5), the underlined values indicating the initial lifespans of their predecessor cells (a pair of sister cells). Shown vertically are the succession of states (or timesteps) traversed by the apical (a_1...i) and basal (b_1...i) cells in the packet during the period from birth (at the level of in-coming arrows) to division (at the level of out-going arrows).

194

arrows which cross-over, like a bootstrap – hence the term bootstrap L-system for this class of algorithm. Notated more formally, the transition rules for P22 are:

$$P = \{a_1 \rightarrow a_2, ..., a_4 \rightarrow a_5, b_1 \rightarrow b_2, ..., b_5 \rightarrow b_6, a_5 \rightarrow b_2a_2, b_6 \rightarrow b_1a_1\}$$

and for P3 they are:

$$P = \{a_1 \rightarrow a_2, ..., a_4 \rightarrow a_5, b_1 \rightarrow b_2, ..., b_3 \rightarrow b_4, a_5 \rightarrow b_1a_1, b_4 \rightarrow b_1a_3\}.$$

By following the transition rules of the two bootstrap systems, two genealogical trees can be constructed (Fig. 11) which represent the temporal sequence of cell division in the packets as they follow their respective pathways in cortex and stele. The trees also show the spatial sequence of division since reading them from left to right at any horizontal level (timestep) shows the number of cells in a basico-apical direction within the packet. The state, or age, of any given cell in the packet at a given timestep is also evident.

The timesteps through which the cells pass have relative units in the notation above. But they need not remain so because the evolution of the packets can be related to real time. The median doubling time for packets in the cortex was estimated as 12 h, and in the stele as 14 h [3]. For the corresponding pathways, P22 and P3, a median of 5 and 4 timesteps per generation was simulated [26]. Therefore, one timestep in each pathway is equivalent to 2.2 h and 3.5 h, respectively. Consequently, cortical cells following P22 have interdivisional periods which range from 8.8 to 13.2 h. In stele cells following P3, the periods range from 10.5 to 17.5 h. This level of variability seems quite acceptable when compared with standard errors estimated for cell cycle duration in the maize root [8].

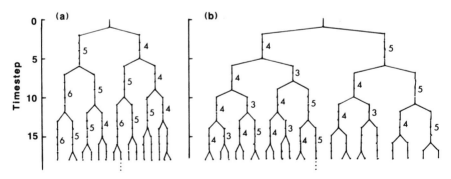

Fig. 11. Genealogical trees of cortical and stelar cell populations, simulated, respectively, by bootstrap devices, which follow (a) division pathway P22 and (b) division pathway P3. The trees make explicit the recurrence graphs of Fig. 10. The nodes on the tree (.) indicate cell states reached at each timestep. Reading horizontally from left to right across the trees at a given timestep shows not only the number of cells in a packet at that step, but also their basico-apical sequence within the packet. The relative lifespans of the cells during each generation are indicated by numbers on the trees.

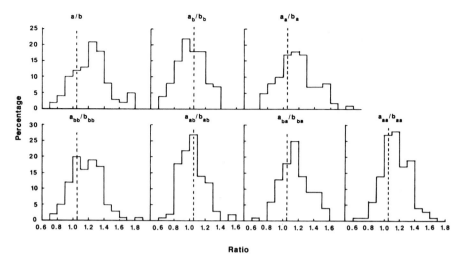

Fig. 12. Histograms showing observed ratios of sister cell lengths for cells within packets of 2, 4, and 8 cells in the cortex of the maize root meristem. Sister cells have either a basal (*b*) or apical (*a*) location in the packet. Lengths of sisters were measured after the completion of different cell generations: two sisters (*b,a*) at the end of generation 1; two pairs of sisters (b_b,a_b) (b_a,a_a) at the end of generation 2; and four pairs of sisters (b_{bb},a_{bb}) (b_{ab},a_{ab}) (b_{ba},a_{ba}) (b_{aa},a_{aa}) at the end of generation 3. The ratios can also be simulated by means of the growth function of the bootstrap system. The best simulation of the observed ratios is given by a combination of pathways P21 and P22.

Each bootstrap system is associated with its own characteristic growth function. This predicts not only the number of cells at a given time, but also the lengths of the cells, assuming that they elongate exponentially. In fact, the ratio, a/b, of sister-cell lengths at each generation can be predicted and compared with those actually found. For example, the region of cortex sampled contains only three pathways, P14, P21 and P22, the latter pair being relatively more frequent than P14 [26]. The a/b ratio of the various pairs of sister cells shown in Fig. 12 is best simulated by a mixture of the two most frequent pathways, P22 + P21 (i.e. (6,5,5,4) and (5,6,5,4)).

The bootstrap systems described above are minimal algorithms specifying the division sequences in the period of packet growth leading to the 8-cell stage. This was the developmental phase most closely analysed. However, further work could undoubtedly devise systems to account for the division sequences leading to the 16-cell stage and beyond. Indeed, any dividing system which shows ancestral dependence of cycle times can be analysed in this way.

5.2. The cytological basis of patterning of proliferative divisions

There are $2!4! = 48$ different division pathways, assuming asynchronous division, by which a single mother cell could theoretically generate an 8-cell packet, and $2!4!8! = 1,935,360$ pathways which lead to a 16-cell packet [26]. In the cortex of the maize root only 3 of the 48 possible pathways leading to the 8-cell packet occurred with significant frequency (>10%), whereas 12 pathways were found in the stele. The greater number of pathways in the stele may reflect the heterogeneity of this cell population due to the development of the many different cell types associated with this tissue.

The relatively few pathways in the cortex suggests that those which are not represented have been excluded, presumably because they cannot be reproduced by the division system of these particular cells. The pathways that are present are actively maintained. The robustness of the particular division pathways is evident from observations on packets in the cortex of roots chilled at 5°C for 4 d and then allowed to recover at 20°C [4]. During the cold period microtubules disperse and divisions cease. Upon re-warming, microtubules reform and divisions recommence [2]. The resumption of division is accompanied by the re-establishment of the pre-existing division patterns, even though there are many other additional pathways that could theoretically be adopted. The same may occur when roots resume division after dormancy has interrupted their embryonic divisional phase [26]. By contrast, growing roots in 5% methanol leads to the eradication of the usual division pathway and the establishment of new ones [4]. This is accompanied by an altered polarity of cytoplasmic staining.

Both sets of experimental results suggest that the division pathways, in general, are programmed by some feature of intracellular organization. Of what this consists is unknown at present, but it may be related to tissue-specific dispositions of the cytoskeleton, such as microtubules and microtubule-organizing centres. The arrangements of cortical microtubules are sensitive to both endogenous growth regulators (e.g., gibberellins) [28] and exogenous metabolic inhibitors, such as those that interfere with RNA synthesis [34]. Thus, it is not unreasonable to suppose that variations in the local environment impinging upon cells within the root could influence, however indirectly, both microtubule dynamics and sites of microtubule formation which in turn could lead to the establishment of tissue-specific patterns of division and cell growth. In fact, the unequal division found in the first cortical cell generation after germination, for example, is likely to result initially in unequal concentrations of tubulin in the two daughter cells. This, in turn, might lead to different microtubule dynamics [19] that influence the relative durations of subsequent interphases. Similar considerations could also apply in connection with the sites of formative divisions within the apex.

6. Significance of formative and proliferative division pathways

Do the division pathways, formative or proliferative, have significance for development? Or, in teleological terms, has the plant made a significant investment in establishing a cellular apparatus capable of maintaining a certain pathway of division? The elegance of both the formative and proliferative division pathways and the L-systems that simulate them (thus implying similarly sophisticated counterparts in the cytological systems) means that the cell cycle should be seen from a new perspective, namely, as one where cells divide not solely as individual entities responding only to patterns of cycle-related gene activity, but as coordinated and organized (programmed) systems in space and time [27].

Pathways of division might also be associated with certain differentiation events. Formative divisions block out groups of cells which eventually form the major tissues. This may be regarded as a *primary* developmental event. Proliferative divisions, with their associated unequal divisions, represent a *secondary* developmental event and provide a fine tuning to the primary event; these divisions may even permit additional cell types to appear within the already blocked-out tissue (e.g. hair and non-hair cells within epidermis) (cf. [16]). The completed organ (the root) then appears as the inevitable outcome of a certain type of general cellular programme of which division is but one aspect. Eventually, it should be possible to formalize the link between the formative division pathways and the subsequent proliferative pathways, thus making explicit a concatenated set of deterministic rules by which large blocks of cells are generated. The pattern of cellularization of these blocks is specific to the species and is thus a phenotypic representation of a certain fraction of the genotype. The genotype also influences the rate of traverse along the division pathways through the well documented relationship between genome mass and cell cycle duration [15, 19]. Whether different genes operate to determine the different classes of division remains to be determined, but in yeast, at least, there are recursive patterns of gene activity [21, 22] associated with division which are reminiscent of the recursive lifespans of proliferative divisions. Also in yeast, there are genes known to affect polarity and division site, some of which interact with *cdc* genes [14].

Although the central dogma [12] dictates that information relevant to development is encoded in DNA and translated into the proteins necessary for the construction of cells, it falls short of specifying how any particular protein is deployed in the construction of organelles such as the mitotic apparatus, for example. However, the mitotic apparatus certainly develops as a result of information stored within its component proteins. An additional type of information, probably stored within the structure of the cell, is needed to specify the pattern of division. Both types of informational system, molecular and cellular, are part of an immensely complex epigenetic code. The L-system mentioned is one way of formalizing this code. It is, however, an 0L-system where there is no interaction between cells. This is a minimal system with which to encode epigenesis. More realistic, maybe, are context-sensitive L-systems which admit interactions between cells, or cell systems that respond to vector fields [13]. These latter can reproduce some of the natural variation in division patterns [5]. At present, work relating cell division to organogenesis is in its infancy. The genetic and epigenetic codes necessary for root (or shoot) organogenesis are likely to be complex and, as Fig. 1 indicates, to be modulated by information derived from many levels of organization which are themselves derived from the operation of these codes. The plant thus represents a system that is essentially an autopoietic loop which drives its own construction [35]. Cell division is obviously central to this process.

Acknowledgement

I am grateful to Drs J. and H.B. Lück for their helpful comments upon this manuscript.

References

1. Armstrong SW and Francis D (1985) Differences in cell cycle duration of sister cells in secondary root meristems of *Cocos nucifera* L. Ann Bot 56: 803–813.
2. Baluska F, Parker JS and Barlow PW (1992) The microtubular cytoskeleton in cells of cold-treated roots of maize (*Zea mays* L.) shows tissue-specific responses Protoplasma 139 (in press).
3. Barlow PW (1987) Cellular packets, cell division and morphogenesis in the primary root meristem of *Zea mays* L. New Phytol 105: 27–56.
4. Barlow PW (1989) Experimental modification of cell division pathways in the root meristem of *Zea mays*. Ann Bot 64: 13–20.
5. Barlow PW (1991) From cell wall networks to algorithms. The simulation and cytology of division patterns in plants. Protoplasma 162: 69–85.
6. Barlow PW (1992) The meristem and quiescent centre in cultured root apices of the *gib-1* mutant of tomato (*Lycopersicon esculentum* Mill.). Ann Bot 69: 533–543.
7. Barlow PW and Adam JS (1989) Experimental control of cellular patterns in the cortex of tomato roots. In: Structural and Functional Aspects of Transport in Roots. BC Loughman, O Gasparíková and J Kolek (eds) pp. 21–24. Dordrecht, Kluwer Academic Publishers.
8. Barlow PW and Macdonald PDM (1973) An analysis of the mitotic cell cycle in the root meristem of *Zea mays*. Proc R Soc Lond, Ser B 183: 385–398.
9. Barlow PW, Parker JS and Brain P (1991) Cellular growth in roots of a gibberellin-deficient mutant of tomato (*Lycopersicon esculentum* Mill.) and its wild-type. J Exp Bot 42: 339–351.
10. Bartlett R and Nurse P (1991) Yeast as a model system for understanding the control of DNA replication in eukaryotes. BioEssays 12: 457–461.
11. Clowes FAL (1954) The promeristem and the minimal constructional centre in grass root apices. New Phytol 53: 108–116.
12. Crick FHC (1958) On protein synthesis. Symp Soc Exp Biol 12: 138–163.
13. De Boer MJM, Fracchia FD and Prusinkiewicz P (1992) A model for cellular development in morphogenetic fields. In: Lindenmayer Memorial Volume. G Rozenberg and A Salomaa (eds) Berlin, Springer-Verlag (in press).
14. Drubin DG (1991) Development of cell polarity in budding yeast. Cell 65: 1093–1096.
15. Francis D, Kidd AD and Bennett MD (1985) DNA replication in relation to DNA C values. In: JA Bryant and D Francis, eds. The Cell Division Cycle in Plants, 61–82. Cambridge University Press, Cambridge.
16. Gunning BES, Hughes JE and Hardham AR (1978) Formative and proliferative cell divisions, cell differentiation, and developmental changes in the meristem of *Azolla* roots. Planta 143: 121–144.
17. Hejnowicz Z (1989) Differential growth resulting in the specification of different types of cellular architecture in root meristems. Env Exp Bot 29: 85–93.
18. Ivanov VB (1971) Critical size of the cell and its transition to division I. Sequence of transition

to mitosis for sister cells in the corn seedling root tip. Sov J Dev Biol 2: 421–428.

19. Ivanov VB (1978) DNA content in the nucleus and and rate of development in plants. Sov J Dev Biol 9: 28–40.

20. Kirschner M and Mitchison T (1986) Beyond self-assembly: from microtubules to morphogenesis. Cell 45: 329–342.

21. Klar AJS (1987) Determinism of yeast cell lineage. Cell 49: 433–435.

22. Klar AJS (1990) The developmental fate of fission yeast cells is determined by the pattern of inheritance of parental and grandparental DNA strands. EMBO J 9: 1407–1415.

23. Lindenmayer A (1984) Models for plant tissue development with cell division orientation regulated by preprophase bands of microtubules. Differentiation 26: 1–10.

24. Lintilhac PM (1987) Plant cytomechanics and its relationship to the development of form. In: Cytomechanics. The Mechanical Basis of Cell Form and Structure. J Bereiter-Hahn, OR Anderson and W-E Reif (eds) pp. 230–241. Berlin, Springer-Verlag.

25. Lloyd CW (1991) How does the cytoskeleton read the laws of geometry in aligning the division plane of plant cells? Development, Suppl 1: 55–65.

26. Lück J, Barlow PW and Lück HB (1992) Cell genealogies in a plant meristem deduced with the aid of a bootstrap L-system (submitted).

27. Lück J and Lück HB (1991) Double-wall cellwork systems for plant meristems. Lect Notes Comput Sci 532: 564–581.

28. Mita T and Katsumi M (1986) Gibberellin control of microtubule arrangement in the mesocotyl epidermal cells of the d5 mutant of *Zea mays* L. Plant Cell Physiol 27: 651–659.

29. Nurse P (1990) Universal control mechanism regulating onset of M-phase. Nature 344: 503–508

30. Potten CS and Loeffler M (1990) Stem cells: attributes, cycles, spirals, pitfalls and uncertainties. Lessons for and from the crypt. Development 110: 1001–1020.

31. Quader H, Wagenbreth I and Robinson DG (1978) Structure, synthesis and orientation of microfibrils. V. On the recovery of *Oocystis solitaria* from microtubule inhibitor treatments. Cytobiologie 18: 39–51.

32. Roland J-C, Reis D, Vian B, Satiat-Jeunemaitre B and Mosiniak M (1987) Morphogenesis of plant cell walls at the supramolecular level: internal geometry and versatility of helicoidal expression. Protoplasma 140: 75–91.

33. Simmonds D, Setterfield G and Brown DL (1983) Organization of microtubules in dividing and elongating cells of *Vicia hajastana* Grossh. in suspension culture. Eur J Cell Biol 32: 59–66.

34. Utrilla L and de la Torre C (1991) Loss of microtubular orientation and impaired development of prophase bands upon inhibition of RNA synthesis in root meristem cells. Plant Cell Reps 9: 492–495.

35. Varela F G, Maturana HR and Uribe R (1975) Autopoiesis: the organization of living systems, its characterization and a model. Biosystems 5: 187–196.

36. Wichmann HE, Loeffler M and Schmitz S (1988) A concept of hemopoietic regulation and its biomathematical realization. Blood Cells 14: 411–429.

15. Regulation of cell division in the shoot apex

DENNIS FRANCIS and ROBERT J. HERBERT

Abstract

The aim of this paper is to explore how changes in the plane, and in the rate, of cell division affect primordium initiation in the shoot apex. Leaf initiation is characterised by a change in the plane of cell division in incipient primordia or changes in the rate of cell division, or both. We refer to published work on the preprophase band (PPB) and the phragmosome to indicate how planes of cell division are predicted in plant cells and argue that the presence of F-actin in PPBs may be a potential substrate for key regulatory cell cycle protein kinases. However the key signalling molecules that may cause a repositioning of PPBs during leaf initiation are unknown. We also attempt to make a link between homeotic, or organ identity genes and the regulation of cell size during floral morphogenesis. Here, emphasis is placed on a model by EM Lord proposing that homeotic genes have heterochronic function. In keeping with a timer mechanism would be genes which regulate cell size at division. The timing mechanism could act first, through homeotic genes determining where and when flower primordia are initiated and second, through genes which cause cell size to alter in cells which are determined as a particular floral primordium.

1. Introduction

The central function of a meristem is to generate new cells by mitotic cell division. In the root meristem, many of the cells produced by cell division will subsequently expand and contribute to the growth of that organ (PW Barlow, this volume). In the shoot meristem, the production of new cells leads to an increase in the size of the apex (that region of the meristem above the youngest primordium, in the case of a distichous or spiral phyllotaxis or youngest primordia in the case of a paired or trimerous, phyllotaxis and often referred to in dicotyledons as the apical dome). Enlargement of the apex occurs until new primordia are initiated on the flanks representing, in effect, the start of the next plastochron. Thus, the apex generates new cells of which only a specific sub-population is destined to form the next leaf primordium. In pea, this was first recognised as a cluster of periclinal divisions on one side of the apical dome which occurred about half-way through the plastochron [17]. In other words, signals are perceived by these cells which then alter their plane of cell division.

J.C. Ormrod and D. Francis (eds.), Molecular and Cell Biology of the Plant Cell Cycle, 201–210.
© 1993 *Kluwer Academic Publishers. Printed in the Netherlands.*

The first aim of this paper is to explore the molecular basis of changes in the plane of cell division and how this may be linked to leaf initiation.

The shoot apex undergoing the floral transition has commanded much attention over the last few years and various review articles have been devoted to it [3, 6, 19, 20]. In particular, cell division has featured as a central part of much of this literature and there seems little point in duplicating these papers. What becomes apparent is that much information now exists on cell division cycle genes *per se* (J Hayles and P Nurse, this volume) and similarly, much new information is available on homeotic genes that direct the development of each whorl of a flower. Moreover, homology exists between on the one hand, central regulatory genes of the cell cycle among quite unrelated organisms (J Hayles and P Nurse; FZ Watts *et al.*, this volume), and on the other, homology between homeotic gene products in plants and those that act as transcription factors in animals and yeasts [29]. The question arises as to the extent to which cell division is linked to the development of a flower. Given the well-defined molecular identity of many of the genes involved in both cell division and floral morphogenesis the second aim of this paper is to explore the possibility of such links.

2. Vegetative development

Changes in the pattern of cell division which bring about leaf initiation in *Pisum sativum* tend to occur in two stages. First, those which cause the change in plane of cell division during the plastochron (see above) and second, those which result in an increase in the rate of cell division which causes the primordium to appear as a bulge on the side of the apex [18].

To understand the nature of the controls which regulate planes of division, elegant studies have been carried out on changes in the arrangement of microtubules in elongating and vacuolating cells during cell division [30]. Epidermal vacuolated cells of *Datura stramonium*, in which the cytoplasm shows clear separation into two large areas as division proceeds, have proved very amenable to such studies. Notably, the nucleus migrates into the centre of the vacuole and is suspended by transvacuolar, cytoplasmic strands [28]. The resultant structure, termed the phragmosome, determines the future division plane. As noted by Flanders *et al.* [5], a better predictor is a cortical ring of microtubules, the pre-prophase band (PPB), which appears in G2 before prophase but which disperses before metaphase is reached [11, 26]. Regardless of the type of division so far investigated, the PPB 'specifies' in which plane the cell will divide. In addition to microtubules which form the PPB, F-actin filaments are present which persist following the depolymerisation of the PPB microtubules [25]. Moreover, studies of isolated preprophase nuclei of roots of *Allium cepa* revealed stable linkages between the nucleus and the PPB [34] and microtubular strands radiate out from the nucleus and contribute to the alignment of the PPB [5]. The transvacuolar cytoplasmic strands are held under tension which, in turn, affect the positioning of the nucleus during the establishment of the division plane [9]. Thus, the plane of cell division is somehow governed by the transient PPB microtubules. Presumably, the association of PPB microtubules with the nuclear membrane results in the generation of signals which are, in turn, linked to the translocation of each set of chromatids in opposite directions. Consistent with this idea would be the contraction of the cortical F-actin into a band at pre-prophase which causes most of the nucleus-associated actin to be formed into the transvacuolar disc, thereby forming the phragmosome within which mitosis and cytokinesis occur [32]. F-actin, resulting from an interaction between G-actin and ATP to give F-actin-ADP, could be a potential substrate within the PPB for an enzyme that facilitates a change in directional growth. The protein kinase activity of p34[cdc2] has been linked to microtubule dynamics [33] and the PPB could be one of its natural substrates in plants cells (see [24] and PCL John *et al.*, this volume). However, actin may have no specific function particularly since not all PPBs contain F-actin [25].

Extrapolation of these model systems to the shoot meristem is not easy since the latter comprises cells which are neither elongating nor are they overtly vacuolated. Nevertheless any signalling molecule(s) which affects the plane of cell division in cells destined to be part of the next primordium presumably act on the positioning of the PPB within each cell. The nature of these signalling molecules is unknown.

Clearly, a change in the plane of cell division is not the sole driving force for primordium initiation. An increase in the rate of cell division often follows for those cells displaying a change in directional growth. Perhaps, the localised increase in cell division activity results from a localised increase in metabolite supply. Moreover the bulging out of the primordium is linked to a change in the plasticity of those epidermal cells which are immediately adjacent to the axis of growth of the primordium [21]. The latter is consistent with a re-orientation of cellulose microfibrils in epidermal cells [27] although how this is regulated is unknown. In other words, an increase in the plasticity of epidermal cell walls

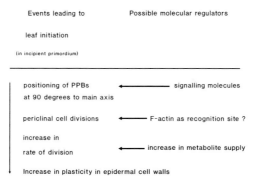

Fig. 1. A scheme linking the activity of hypothetical signalling molecules to the positioning of preprophase bands (PPB) in relation to leaf initiation.

accommodates the change in polarity of growth of the underlying cells and the net effect is a bulge on the side of the apex (Fig. 1). However, this somewhat simplistic view must be tempered by observations that show interspecific differences in the events that occur during leaf initiation. For example, in *Silene coeli-rosa* the occurrence of periclinal cell division bears little relationship to the timing of the initiation of the pair of primordia that form at the start of each plastochron. Lyndon and Cunninghame [21] concluded that changes in the plane of cell division and changes in the rate of cell division may be 'permissive' rather than 'causal' factors in leaf initiation. Another factor must surely be interspecific variation in apical structure and phyllotaxis where the geometry of the apex will dictate the relative importance of changes in direction of growth and changes in the plane of cell division. Such discussions also highlight the paucity of data that exist on leaf initiation where it is difficult to make generalisations about mechanisms when available data allow us to examine leaf initiation in only a handful of angiosperm species.

3. Floral development

An enlargement of the shoot apex occurs before the initiation of the floral whorls in a number of species [20]. This increased rate of growth is co-incidental with a synchronisation of cell division in two long day plants, *Silene coeli-rosa* [7] and *Sinapis alba* [1]. Synchrony has been discussed extensively [6, 22] and given that it does not occur in short day plants such as *Xanthium strumarium* [15] and *Pharbitis nil* [12], it may only be related to the initiation of floral parts in some species. Other changes in the cell cycle have been recorded at those times when the apex is becoming committed to flower, a phase known as floral evocation [4]. Marked differences in the pattern of alteration to the component phases of the cell cycle during floral evocation co-incide with differences at the molecular level. For example, the synchronisation of the cell cycle in the apical dome of *Silene* (see above) occurs at the same time as the appearance of new polypeptides [8]. *Silene* requires 7 long days (experimental days 0-6) for 100% flowering but only on day 8 have qualitative changes in polypeptide composition been detected [8]. Paradoxically, we suggested that the change in the cell cycle which leads to synchrony was a marker of underlying molecular changes. Subsequently, a treatment which suppresses synchrony (7LD + 48 h darkness [10]) suppressed neither the qualitative change in polypeptide composition nor flowering [31].

In the short day plant, *Pharbitis nil*, an inductive treatment (48 h darkness) given to four-day-old seedlings resulted in a maximal difference in both the mean cell doubling time [13] and the cell cycle in the terminal shoot apex at the transition from darkness to light compared with plants given a non-inductive treatment (48 h darkness interrupted with red light at 8 and 32 h). In the inductive treatment, G2-phase was eliminated almost entirely in a cell cycle of 22 h in the peripheral zone, whereas there was a six-fold increase in the duration of G2 in a 40 h cell cycle in the peripheral zone in the non-inductive treatment [12]. More recent work using a PCR amplification of cDNA prepared from total RNA from the apical domes at this time, followed by differential screening of the cDNA libraries, has revealed inserts in the floral library which hybridise exclusively with a floral probe prepared from the original cDNA. Following primary, secondary and tertiary differential screening, four clones from a sub-selection of the floral cDNA library hybridise exclusively to a floral probe (RJ Herbert, D Francis, K Edwards, unpublished). In other words, here is an example where a change to the cell cycle as a result of an inductive treatment does reflect an underlying change in gene expression which has yielded at least four putative novel clones. Further analysis of these floral-related clones is in progress. Molecular changes which coincide with marked changes to the cell cycle may represent the expression of regulatory genes that facilitate the changes which then lead to floral morphogenesis. Moreover, any floral specific gene expression occurring during floral evocation is some way ahead of the expression of homeotic, or organ identity, genes (see below).

A shortened cell cycle is a prerequisite for flowering in both long day and short day plants [20]. The well-worked examples include the long day plants *Silene coeli-rosa* and *Sinapis alba*. In *Silene*, the terminal apex results in a single terminal flower and, subsequently, the axillary buds flower to produce a cymaceous inflorescence [23]. In *Sinapis*, the vegetative apex transforms into an inflorescence. Thus, primordia are initiated as before but the axillary part of each one grows out to form a floral shoot. The major advances on homeotic genes (organ identity genes) have come from studying plants where the indeterminate inflorescence apex is not converted into a floral meristem (e.g. *Antirrhinum* and *Arabidopsis*). In *Antirrhinum*, plant form is controlled by three homologous meristems: vegetative, inflorescence and floral while in *Arabidopsis*, the spatial arrangement of the three meristems is different [2].

What is clear is that cell division studies have concentrated on plant systems where floral morphogenesis is a simple transition from a vegetative to a floral meristem. Moreover the transition to flowering can be controlled very tightly by photoperiod and cell division is studies in homogenous batches of plants. However, the identification of homeotic genes has been achieved in plants which are characterised genetically, have well-characterised transposable elements but where flowering cannot be tightly controlled by photoperiod. Hence, very little is known about the changes in cell division which occur during floral morphogenesis in *Antirrhinum* and *Arabidopsis*. It may be worthwhile to explore what the link could be between cell division activity during floral evocation and floral morphogenesis in relation to the expression of organ identity genes.

Various mutants of *Arabidopsis* and *Antirrhinum* exist where the wrong organ develops in the wrong place or where an organ fails to develop or fails to develop normally. These abnormalities are a consequence of a mutation in a gene that affects differentiation. Sommer *et al.* [29] assigned mutations of homeotic genes that specify organ identity to three different categories:

- type 1, 1st and 2nd whorl organs (perianth) are affected
- type 2, 3rd and 4th whorl organs are affected
- type 3, 2nd and 3rd whorls are altered

Organ identity genes in *Arabidopsis thaliana* and *Antirrhinum majus* may encode transcription factors. The organ identity genes, *agamous* (*AG* in *Arabidopsis*) and *deficiens* (*DEF* in *Antirrhinum*), encode proteins which show high degrees of sequence homology to the conserved DNA binding domains of two known transcription factors in animals and yeasts, *serum response factor* (*SRF*) and *mini chromosome maintenance protein* (*MCM*1), respectively. By taking the first letter of each of these gene products, Sommer *et al.* [26] used the term *MADS*-box to denote this group of proteins.

What is not known are the factors that signal the onset of flowering facilitating the expression of MADS-box genes. Hence it will be important to resolve amino-acid sequences that are encoded by floral-specific genes isolated during floral evocation (see above). In turn, the genes or proteins that the transcription factors act on to bring about the initiation of individual organs of the flower are unknown. Recently Lord [16] used one of the homeotic mutants of *Arabidopsis* (*pistillata*) [14] to illustrate a temporal model of organ initiation followed by organ differentiation. In this model, the *pistillata* gene is not necessary for petal initiation but only controls the later stages of petal development. In the mutant, a mosaic organ forms with both petal and sepal features and may result from overlapping programmes operating at one point in time. The model accepts the premise that an, as yet undiscovered, set of genes is responsible for primordial initiation in a specific floral formula. These genes then induce the organ identity genes which then assign a specific identity to each set of organs. Lord concluded 'there would have to be a perfect alignment between the genes controlling the time of primordial initiation and the identity gene 'cascade' to get a normal flower'. The fact that hybrid organs are produced in the homeotic mutants suggests that this alignment has been disrupted.

The concept of temporal controls on differentiation is a somewhat neglected aspect of plant development. Whilst it is clear that cells at a specific position on the apex become part of 'the next primordium' timing becomes critical in determining when a particular primordium develops. One way of interpreting a homeotic mutant is that the timing of primordium initiation is perturbed so that a whorl is missed out or the time taken to differentiate into a particular floral primordium is perturbed [14].

A link between timing of primordium initiation and cell division becomes obvious if we assume that cells grow to a particular size before dividing (see PCL John *et al.*, this volume). The evidence for such controls in fission yeast is clear with the genes *nim*1, *wee*1 and *cdc*25 acting on the p34 protein kinase [see J Hayles and P Nurse, this volume]. Given the homology that exists between these genes among a wide range of unrelated organisms, a timing mechanism must surely operate on cell division in plants.

In our model, the timer is a heterochronic gene which initiates a cascade of

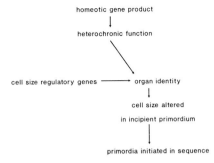

Fig. 2. A scheme proposing a heterochronic function for homeotic genes in relation to the initiation of floral primordia.

changes that regulates when primordium initiation begins (Fig. 2). The heterochronic function would be well ahead of the activity of the genes which regulate cell size. It would be a case of the regulatory cell size genes acting on a particular cohort which are determined as a particular primordium. For example, the cells in an incipient floral primordium may divide at a smaller cell size than adjacent cells. If so, altered sizer controls would, in effect, distinguish the primordium cells as a separate domain from surrounding cells. Our model argues for an interaction between homeotic genes and cell division cycle genes in that a temporal control will operate through cell division in order that floral primordia are initiated in a normal rhythmic sequence. Lyndon and Cunninghame [21] found that in *Silene*, cell doubling times oscillate between faster rates and then shorter rates as successive various whorls are initiated. Thus, it may be expected that cell size at division is different in each whorl. Preliminary results from *Arabidopsis* suggest this is so at least for petals compared with other floral organs (W Crone and E Lord, unpublished). Perturbations to sizer controls may also disrupt the normal sequence of events and would help to explain why in some homeotic mutants, primordia of a particular whorl fail to develop, or are replaced by abnormal organs. Clearly, data on cell size in incipient floral primordia would be required to test these hypotheses.

4. Conclusions

1. Given the advances made on the resolution of genes which affect cell division and flower development it should be possible to seek answers to the difficult question – what determines shape?

2. A change in the plane of cell division is an important aspect of leaf initiation although the extent to which it is a primary event is dependent on the species in question. The establishment of periclinal divisions in regions of the apex which were, hitherto, dominated by anticlinal cell division establishes a

change in the direction of growth for an incipient primordium. This has been well-documented and is not a new idea of ours. Based on the model systems used to show that the positioning of the pre-prophase band influences the plane of a cell division we argue that the PPB or its vestigial remnants (F-actin) serve as one of the substrates for signalling molecules which promote a change in directional growth. We accept that F-actin may not be a universal component of PPB's but suggest that *in situ* hybridisation at the protein level, deploying a monoclonal antibody to the conserved PSTAIR sequence of p34 [35], should indicate whether the plant homologue of p34^{cdc2} binds in the vicinity of the PPB. We also accept that p34^{cdc2} probably has pleiotropic effects on the initiation of cell division which would not make for simple interpretations of *in situ* experiments.

3. Floral development can be interpreted as a function of time so that homeotic or organ identity genes take on a heterochronic function or may themselves be controlled by an earlier acting set of heterochronic genes. We emphasise that this model is not ours but was first proposed by Lord [16]. We have simply added to this model by proposing that genes which regulate cell size contribute to the timing mechanism by altering cell size at division for the cells of incipient floral primordia.

In this article, the emphasis has been to generate hypotheses which may stimulate others to take up the challenge of asking questions about cell division in relation to shoot development. Given the technology now at our disposal, and elegantly displayed throughout this volume, answers should be forthcoming.

Acknowledgements

Unpublished work mentioned in this article was supported by a SERC biotechnology CASE studentship to RJH and we thank ICI plc for additional financial support and Professor Elizabeth Lord (UC Riverside) for critically reviewing the manuscript.

References

1. Bernier G, Kinet J-M and Bronchart R (1967) Cellular events at the meristem during floral evocation in *Sinapis alba* L. Physiol Veg 5: 311-324.
2. Coen ES (1991) The role of homeotic genes in flower development and evolution. Annu Rev Pl Physiol Pl Mol Biol 42: 241-279.
3. Coen ES and Meyerowitz EM (1991) The war of the whorls: genetic interactions controlling flower development. Nature 353: 31-37.
4. Evans LT (1971) The nature of floral induction. In: The Induction of Flowering. LT Evans (ed), pp. 457-480. Melbourne: Macmillan.

5. Flanders DJ, Rawlins DJ, Shaw PJ and Lloyd CW (1990) Nucleus-associated microtubules help determine the division plane of plant epidermal cells: Avoidance of four-way junctions and the role of cell geometry. J Cell Biol 110: 1111-1122.

6. Francis D (1991) The cell cycle in plant development. Tansley Rev No. 38 New Phytol 122: 1–20.

7. Francis D and Lyndon RF (1979) Synchronisation of cell division in the shoot apex of *Silene* in relation to flower initiation. Planta 145: 151-157.

8. Francis D, Rembur J and Nougarède A (1988) Changements dans la composition polypeptidique du méristème de *Silene coeli-rosa* (L.) au cours de l'induction florale. Comptes Rendues Acad Sci Paris Ser III 307: 763-770.

9. Goodbody KC, Venverloo J and Lloyd CW (1991) Laser microsurgery demonstrates that cytoplasmic strands anchoring the nucleus across the vacuole of premitotic cells are under tension. Implications for division plane alignment. Development 113: 931-939.

10. Grose S and Lyndon RF (1984) Inhibition of growth and synchronised cell division in the shoot apex in relation to flowering in *Silene*. Planta 161: 289-294.

11. Gunning BES (1982) The cytokinetic apparatus: its developmental and spatial regulation. In: The Cytoskeleton in Plant Growth and Development. CW Lloyd (ed), pp. 229-292. New York: Academic Press Inc.

12. Herbert RJ (1992) Cellular and molecular studies on the shoot terminal meristem of *Pharbitis nil* Chois. cv. Violet during floral evocation Ph D Thesis, Univ Wales.

13. Herbert RJ, Francis D and Ormrod JC (1992) Cellular and morphological changes at the terminal shoot apex of the short day plant *Pharbitis nil* Chois. cv. Violet, during the transition to flowering. Physiol Pl 86: 85–92.

14. Hill JP and Lord EM (1989) Floral development in *Arabidopsis thaliana*: a comparison of the wild type and the homeotic *pistillata* mutant. Can J Bot 67: 2922-2936.

15. Jacqmard A, Raju MVS, Kinet J-M and Bernier G (1976) The early action of the floral stimulus on mitotic activity and DNA synthesis in the apical meristem of *Xanthium strumarium*. Am J Bot 63: 166-174.

16. Lord EM (1991) The concepts of heterochrony and homeosis in the study of floral morphogenesis. In: G Bernier (ed.) Flowering Newslett 11: 4-13.

17. Lyndon RF (1970) Planes of cell division and growth in the shoot apex of *Pisum*. Ann Bot 34: 19-28.

18. Lyndon RF (1983) The mechanism of leaf initiation. In: The Growth and Functioning of Leaves. JE Dale and FL Milthorpe (eds), pp. 3-24. Cambridge: Cambridge Univ Press.

19. Lyndon RF (1990) Plant Development. The Cellular Basis. London, New York: Unwin Hyman Inc.

20. Lyndon RF and Francis D (1984) The response of the shoot apex to light-generated signals from the leaves. In: Light and the Flowering Process. D Vince-Prue B Thomas and KE Cockshull (eds), pp. 171-192. London: Acad Press.

21. Lyndon RF and Cunninghame ME (1986) Control of shoot apical development via cell division. In: Plasticity in Plants, 40th Sym Soc Exp Biol. AJ Trewavas and DH Jennings (eds), pp. 233-255. Cambridge: Company of Biologists.

22. Lyndon RF and Francis D (1992) Plant and organ development. Plant Mol Biol 19: 51-68.

23. Miller MB and Lyndon RF (1976) Rates of cell division in the shoot apex of *Silene* during the transition to flowering. J Exp Bot 27: 1142-1153.

24. Mineyuki Y, Yamashita M and Nagahama Y (1991) p34^{cdc2} kinase homologue in the preprophase band. Protoplasma 162: 182-186.

25. Palevitz BA (1987) Actin in the preprophase band of *Allium cepa*. J Cell Biol 104: 1515-1519.

26. Pickett-Heaps JD and Northcote DH (1966) Organisation of microtubules and endoplasmic reticulum during mitosis and cytokinesis in wheat meristems. J Cell Sci 1: 109-120.

27. Selker JML and Green PB (1984) Organogenesis in *Graptopetalatum paraguayense* E. Walther: shifts in orientation of cortical microtubule arrays are associated with periclinal divisions. Planta 160: 289-297.

28. Sinnott E and Bloch R (1940) Cytoplasmic behaviour during division in vacuolate plant cells. Proc Natl Acad Sci USA 26: 223-227.
29. Sommer H, Beltran J, Huijser, Pape H, Loning W, Saedler H and Schwarz-Sommer Z (1990) *Deficiens*, a homeotic gene involved in the control of flower morphogenesis in *Antirrhinum majus*: the protein shows homology to transcription factors. EMBO J 9: 605-613.
30. Staiger CJ and Lloyd CW (1991) The plant cytoskeleton. Curr Opinion Cell Biol 3: 33-42.
31. Taylor M, Francis D, Rembur J and Nougarède A (1990) Changes to proteins in the shoot meristem of *Silene coeli-rosa* during the transition to flowering. Plant Cell Physiol 31: 1169-1176.
32. Traas JA, Doonan JH, Rawlins DJ, Shaw PJ, Watts J and Lloyd CW (1987) An actin network is present in the cytoplasm throughout the cell cycle of carrot cells and associates with the dividing nucleus. J Cell Biol 105: 387-395.
33. Verde F, Labbé JC, Dorée M and Karseti E (1990) Regulation of microtubule dynamics by *cdc*2 protein kinase in cell-free extracts of *Xenopus* eggs. Nature 343: 233-238.
34. Wick SM and Duniec J (1983) Immunofluorescence microscopy of tubulin and microtubule arrays in plant cells. I. Preprophase band development and concomitant appearance of nuclear envelope-associated tubulin. J Cell Biol 97: 235-243.
35. Yamashita M, Yoshikuni M, Hirai T, Fukada S and Nagahama Y (1991) A monoclonal antibody against the PSTAIR sequence of p34[cdc2], catalytic subunit of maturation-promotion factor and key regulator of the cell cycle. Develop Growth & Differ 33: 617-624.

List of contributors

S.J. AVES
Department of Biological Sciences, University of Exeter, Washington-Singer Laboratories, Perry Road, Exeter, EX4 4Q, U.K.

P.W. BARLOW
Department of Agricultural Sciences, University of Bristol, AFRC Institute of Arable Crops Research, Long Ashton Research Station, Long Ashton, Bristol, BS18 9AF, U.K.

L. BAKÓ
Institute of Plant Physiology, Biological Research Center, Hungarian Academy of Sciences, Szeged, P.O. Box 521, 6701 Hungary

L. BÖGRE
Institute of Plant Physiology, Biological Research Center, Hungarian Academy of Sciences, Szeged, P.O. Box 521, 6701 Hungary

C.M. BRAY
Department of Biochemistry and Molecular Biology, University of Manchester, Oxford Road, Manchester, M13 9PT, U.K.

J.A. BRYANT
Department of Biological Sciences, University of Exeter, Washington Singer Laborator, Perry Road, Exeter, EX4 4Q, U.K.

M.R.H. BUDDLES
Department of Biochemistry and Molecular Biology, University of Manchester, Oxford Road, Manchester, M13 9PT, U.K.

S.R. BURGESS
Department of Agricultural Sciences, University of Bristol, AFRC Institute of Arable Crops Research, Long Ashton Research Station, Long Ashton, B ol, BS18 9AF, U.K.

J.F. BURKE
School of Biological Sciences, University of Sussex, Falmer, Brighton, BN1 9QG, U.K.

N.J. BUTT
School of Biological Sciences, University of Sussex, Falmer, Brighton, BN1 9QG, U.K.

W.Z. CANDE
Department of Molecular and Cellular Biology, University of California, Berkeley, U.S.A.

J.V. CARBAJOSA
Departamento de Bioquímica y Biología Molecular, Escuela Técnica Superior de Ingenieros Agrónomos, Universidad Politécnica de Madrid, 2804 Madrid, Spain

D. CHIATANTE
Sezione di Botanica Generale, Dipartimento di Biologia, Università degli Studi di Milano, via Celoria 26, Milano, Italy

A. CLARKE
School of Biological Sciences, University of Sussex, Falmer, Brighton, BN1 9QG, U.K.

L. COMAI
Department of Botany, University of California, Davis, CA 95616-8517, U.S.A.

D. DEDEOGLU
Institute of Plant Physiology, Biological Research Center, Hungarian Academy of Sciences, Szeged, P.O. Box 521, 6701 Hungary

S. DEL DUCA
Dipartimento di Biolgia e.s., Sede di Botanica, Via Irnerio 42, 40136 Bologna, Italy

C. DONG
Plant Cell Biology Group, Research School of Biological Sciences, Australian National University, Canberra, A.C.T. 2601, Australia

D. DUDITS
Institute of Plant Physiology, Biological Research Center, Hungarian Academy of Sciences, Szeged, P.O. Box 521, 6701 Hungary

F. FELFÖLDI
Institute of Plant Physiology, Biological Research Center, Hungarian Academy of Sciences, Szeged, P.O. Box 521, 6701 Hungary

D. FRANCIS
School of Pure and Applied Biology, University of Wales, College of Cardiff,
P.O. Box 915, Cardiff CF1 3TL, U.K.

J. GYÖRGYEY
Institute of Plant Physiology, Biological Research Center, Hungarian
Academy of Sciences, Szeged, P.O. Box 521, 6701 Hungary

N.G. HALFORD
Department of Agricultural Sciences, University of Bristol, AFRC Institute of
Arable Crops Research, Long Ashton Research Station, Long Ashton, Bristol,
BS18 9AF, U.K.

M.J. HAMER
Department of Biochemistry and Molecular Biology, University of
Manchester, Oxford Road, Manchester, M13 9PT, U.K.

D.E. HANKE
University of Cambridge, Department of Plant Sciences, Downing Street,
Cambridge, CB2 3EA, U.K.

J. HAYLES
Cell Cycle Group, University of Oxford, South Parks Road, OX1 3QU, U.K.

T. KAPROS
Institute of Plant Physiology, Biological Research Center, Hungarian
Academy of Sciences, Szeged, P.O. Box 521, 6701 Hungary

P.C.L. JOHN
Plant Cell Biology Group, Research School of Biological Sciences, Australian
National University, Canberra, A.C.T. 2601, Australia
The Cooperative Research Centre for Plant Science, c/o RSBS, G.P.O. Box 4,
A.C.T. 2601, Australia

P. LAYFIELD
School of Biological Sciences, University of Sussex, Falmer, Brighton, BN1
9QG, U.K.

J.S. MACHUKA
School of Biological Sciences, University of Sussex, Falmer, Brighton, BN1
9QG, U.K.

Z. MAGYAR
Institute of Plant Physiology, Biological Research Center, Hungarian
Academy of Sciences, Szeged, P.O. Box 521, 6701 Hungary

A.L. MOORE
Cell Cycle Group, University of Oxford, South Parks Road, 0X1 3QU, U.K.

T. NAGATA
Department of Biology, Faculty of Science, University of Tokyo, Hongo, Bunkyo-ku, Tokyo 113, Japan

P. NURSE
Cell Cycle Group, University of Oxford, South Parks Road, OX1 3QU, U.K.

J.C. ORMROD
I.C.I. Agrochemicals, Jealott's Hill Research Station, Bracknell, U.K.

J.S. PARKER
Department of Agricultural Sciences, University of Bristol, AFRC Institute of Arable Crops Research, Long Ashton Research Station, Long Ashton, Bristol, BS18 9AF, U.K.

J. ROSAMOND
Department of Biochemistry and Molecular Biology, University of Manchester, Oxford Road, Manchester, M13 9PT, U.K.

T.L. ROST
Department of Botany, University of California, Davis, CA 95616-8517, U.S.A.

P.A. SABELLI
Department of Agricultural Sciences, University of Bristol, AFRC Institute of Arable Crops Research, Long Ashton Research Station, Long Ashton, Bristol, BS18 9AF, U.K.

D. SERAFINI-FRACASSINI
Dipartimento di Biolgia e.s., Sede di Botanica, Via Irnerio 42, 40136 Bologna, Italy

P.R. SHEWRY
Department of Agricultural Sciences, University of Bristol, AFRC Institute of Arable Crops Research, Long Ashton Research Station, Long Ashton, Bristol, BS18 9AF, U.K.

C.J. STAIGER
Department of Cell Biology, John Innes Centre for Plant Science Research, Norwich, NR4 7UH, U.K.

E.Y. TANIMOTO
Department of Botany, University of California, Davis, CA 95616-8517, U.S.A.

F.Z. WATTS
School of Biological Sciences, University of Sussex, Falmer, Brighton, BN1 9QG, U.K.

K. ZHANG
Plant Cell Biology Group, Research School of Biological Sciences, Australian National University, Canberra, A.C.T. 2601, Australia

Subject Index